Electronics

FOR

DUMMIES®

2ND EDITION

Electronics
FOR
DUMMIES®
2ND EDITION

by Cathleen Shamieh and Gordon McComb

WILEY

Wiley Publishing, Inc.

Electronics For Dummies®, 2nd Edition

Published by
Wiley Publishing, Inc.
111 River Street
Hoboken, NJ 07030-5774
www.wiley.com

For general information on our other products and services, please contact our Customer Care Department within the U.S. at 877-762-2974, outside the U.S. at 317-572-3993, or fax 317-572-4002.

For technical support, please visit www.wiley.com/techsupport.

Wiley also publishes its books in a variety of electronic formats. Some content that appears in print may not be available in electronic books.

Library of Congress Control Number: 2009933743

ISBN: 978-0-470-28697-5

Manufactured in the United States of America

10 9 8 7 6 5 4 3 2

WILEY

About the Authors

Cathleen Shamieh is a writer with an engineering background who special-izes in creating communication materials focused on the business benefits of technology. She received an outstanding education in electrical engineer-ing at Manhattan College and MIT, and worked as an engineer in the medical electronics and telecommunications industries before shifting her career into marketing communications and business consulting for high-tech companies. Cathleen enjoys leveraging her technical and business background to create white papers and other materials for not-so-technical audiences.

Gordon McComb has penned more than 60 books and over 1,000 magazine articles. More than a million copies of his books are in print, in over a dozen languages. For 13 years, Gordon wrote a weekly syndicated newspaper column on personal computers. When not writing about hobby electronics and other fun topics, he serves as a consultant on digital cinema to several notable Hollywood clients.

Dedication

To my parents, Beth and Jim Corbett, who taught me that I can do anything I put my mind to; to Sister Eustelle, who made a writer out of me; to my wonderful husband, Bill, who's always there to support me; and to my four fantastic sons, Kevin, Peter, Brendan, and Patrick, who make life a fun, loving adventure every single day.

C. S.

To my father, Wally McComb, who instilled in me a fascination with electronics; and to Forrest Mims, who taught me a thing or two about it.

G. M.

Authors' Acknowledgments

Cathleen Shamieh extends her thanks to the excellent editors at Wiley, especially Katie Feltman and Christopher Morris, for their hard work, support, and gentle reminders, and to Kirk Kleinschmidt, for his intense technical scrutiny of the material. She is also grateful to Linda Hammer and Ken Donoghue, who kindly recommended her work to Wiley. Finally, Cathleen thanks her family and friends, whose support, assistance, and understanding helped make her goal of becoming a *For Dummies* author a reality.

Author Gordon wishes to thank his family, who once again put their lives on hold while he finished another book.

Publisher's Acknowledgments

We're proud of this book; please send us your comments through our online registration form located at `http://dummies.custhelp.com`. For other comments, please contact our Customer Care Department within the U.S. at 877-762-2974, outside the U.S. at 317-572-3993, or fax 317-572-4002.

Some of the people who helped bring this book to market include the following:

Acquisitions, Editorial

Sr. Project Editor: Christopher Morris

Acquisitions Editor: Katie Feltman

Sr. Copy Editor: Barry Childs-Helton

Technical Editor: Kirk Kleinschmidt

Editorial Manager: Kevin Kirschner

Sr. Editorial Assistant: Cherie Case

Cartoons: Rich Tennant
(`www.the5thwave.com`)

Composition Services

Project Coordinator: Katie Crocker

Layout and Graphics: Karl Brandt, Shawn Frazier, SDJumper

Proofreader: John Greenough

Indexer: Potomac Indexing

Publishing and Editorial for Technology Dummies

  Richard Swadley, Vice President and Executive Group Publisher

  Andy Cummings, Vice President and Publisher

  Mary Bednarek, Executive Acquisitions Director

  Mary C. Corder, Editorial Director

Publishing for Consumer Dummies

  Diane Graves Steele, Vice President and Publisher

Composition Services

  Debbie Stailey, Director of Composition Services

Contents at a Glance

Table of Contents

Introduction

. . . .

Are you curious to know what makes your iPod tick? How about your cellphone, laptop, stereo system, digital camera, 46-inch plasma TV — well, just about every other electronic thing you use to entertain yourself and enrich your life?

If you've ever wondered how transistors, capacitors, and other building blocks of electronics work, or if you've been tempted to try building your own electronic devices, you've come to the right place!

Electronics For Dummies, 2nd Edition, is your entrée into the electrifying world of modern electronics. No dry, boring, or incomprehensible tome, this; what you hold in your hands is *the* book that enables you to understand, create, and troubleshoot your own electronic devices.

Why Buy This Book?

All too often, electronics seems like a mystery because it involves controlling something you can't see — electric current — which you've been warned repeatedly not to touch. That's enough to scare away most people. But as you continue to experience the daily benefits of electronics, you may begin to wonder how it's possible to make so many incredible things happen in so many small spaces.

This book is designed to explain electronics in ways you can relate to. It gives you a basic understanding of exactly what electronics is, offers down-to-earth explanations of how major electronic components work, and provides just what you need to build and test working electronic circuits and projects. Although this book doesn't pretend to answer all your questions about electronics, it gives you a good grounding in the essentials.

It is our hope that when you're done with this book, you realize that electronics really isn't as complicated as you may have once thought. And, it is our intent to arm you with the knowledge and confidence you need to charge ahead in the exciting field of electronics.

Why Electronics?

Electronics is everywhere. You find electronics in your communication devices, entertainment systems, and kitchen appliances. Electronic systems control traffic lights, Internet commerce, medical devices — even many toys. Try for just one minute to imagine your life without electronics — you might as well be living in the Dark Ages!

So, what does all this mean to you as you peruse this book? After all, you don't expect to be able to design satellite communication systems after a sit-down session with this humble *For Dummies* book. Although that statement is true, it's also true that even the most complicated electronic systems consist of no more than a handful of different electronic component types governed by the same set of rules that determine the functionality of simple circuits. So, if you want to glean an understanding of complex electronic systems, you start with the basics — just like the designers of those systems did when they got started.

More importantly, understanding the basics of electronics can enable you to create some truly useful, albeit simple, electronic devices. You can build circuits that flash lights at just the right time, sound a buzzer upon sensing an intruder, or even move an object around the room. And, when you know how to use integrated circuit (IC) chips, which are populated with easy-to-use, fully functioning miniaturized circuits, you can create some rather involved designs that will impress your friends and enemies — for just a few well-spent bucks.

With technology development being what it is — lightning fast and smaller and less expensive year after year — you can now hold the ingredients for advanced electronic systems in the palm of your hand. With a little knowledge and a willingness to experiment, you can build something that controls the lighting in your entire house, a robot that vacuums your living room, or an alarm system that senses someone trying to open your refrigerator.

You may have another hobby that can be enriched by your knowledge of electronics. If you're into model railroading, for example, you can use your knowledge of electronics to build your own automated track switchers. If your hobby is racing radio-controlled cars, electronics know-how may enable you to improve the performance of your car and beat your best friend in the next race.

Last, but not least, electronics is fun. Gaining knowledge about and experiencing electronics is its own reward.

Foolish Assumptions

This book assumes that you're curious about electronics but don't know much, if anything, about its inner workings. Because you chose this book, rather than a book consisting exclusively of recipes for electronic circuits, we assume that you want to find out more about how parts such as resistors, capacitors, and transistors actually work, so we take the time (and more than half this book) to explain it to you, distilling fairly technical information into easy-to-understand concepts. You don't need to be well versed in physics or mathematics to benefit from reading this book, although a teeny bit of high school algebra would be helpful (but we do our best to refresh that possibly painful memory).

We assume that you may want to jump around this book a bit, diving deep into a topic or two that holds special interest for you and possibly skimming through other topics. For this reason, we provide loads of chapter cross-references to point you to information that can fill in any gaps or refresh your memory on a topic. And, though the first half of this book is devoted to how electronic circuits and individual parts work, we include cross-references to learning circuits and projects that appear later in the book. That way, as soon as you understand a component, you can jump ahead, if you want, and build a circuit that uses that component.

The table of contents at the front of this book provides an excellent resource that you can use to quickly locate exactly what you're looking for. You'll also find the glossary useful when you get stuck on a particular term and need to review its definition. Finally, the folks at Wiley have thoughtfully provided a thorough index at the back of the book to assist you in narrowing your reading to specific pages.

Safety Is Number 1

Reading about electronics is safe. Probably the worst thing that can happen is that your eyes grow tired from too many late nights spent reading this book. *Building* electronic projects is another matter, though. Lurking behind the fun of your electronics hobby are high voltages that can electrocute you, soldering irons that can burn you, and little bits of wire that can fly into your eyes when you snip them off with sharp cutters. Ouch!

Safety is *numero uno* in electronics. It's so important, in fact, that we devote a major section of Chapter 9 to it — and continually refer you to that section.

If you're brand-new to electronics, please be sure to read the section. Don't skip over it, even if you think you're the safest person on earth. Even if you've dabbled in electronics, it never hurts to refresh your safety memory. When you follow proper precautions, electronics is an extremely safe and sane hobby. Be sure to keep it that way!

Although we try to give you helpful advice about safety throughout, we can't possibly give you, in one book, every possible safety precaution. In addition to reading our advice, use your own common sense, read manufacturer instructions for parts and tools you work with, and always stay alert.

How This Book Is Organized

Electronics For Dummies is organized so that you can quickly find, read, and understand the information you want. The book is also organized so that if you have some experience with electronics, or want to deepen your knowledge of a particular topic, you can skip around and focus on the chapters that interest you.

The chapters in this book are divided into parts to help you zero in, quickly and easily, on the information you're looking for.

Part I: Understanding the Fundamentals of Electronics

Turn to Part I if you want to get a thorough grounding in basic electronics theory. Chapter 1 gives you the "big picture" of exactly what electronics is and the amazing things it can do for you. You discover the fundamentals of electronic circuits and are introduced to voltage, current, and sources of electrical energy in Chapter 2. In Chapters 3 through 6, you dive deep into the heart of all the major electronic components, including resistors, capacitors, inductors, transformers, diodes, and transistors. You find out how each component works, how it handles electric current, and what role it plays in electronic circuits. Chapter 7 introduces you to integrated circuits (ICs) and explains a bit about digital logic and how three popular ICs function. Chapter 8 covers sensors, speakers, buzzers, switches, wires, and connectors. Throughout Part I, we point you to introductory circuits you can build in Part III to demonstrate the operation of each component.

Part II: Getting Your Hands Dirty

Part II is geared around "tooling up," constructing real circuits, and probing around working (and nonworking) circuits — while warding off electrocution. In Chapter 9, you find out how to set up an electronics workbench, which electronic components, tools, and other supplies you need in order to build circuits, and how to protect yourself and your electronic components as you work on circuits. Chapter 10 explains how to interpret circuit diagrams (known as *schematics*) so that you know how to connect components when you build a circuit. You explore various methods of wiring up temporary and permanents circuits in Chapter 11, which also instructs you in the ways of soldering. Finally, Chapters 12 and 13 explain how to use three of the most important testing tools in electronics — the multimeter, logic probe, and oscilloscope — to explore and analyze circuit behavior.

Part III: Putting Theory into Practice

If you're eager to wire up some circuits and get your electronic juices flowing, Part III is the place to be. Chapter 14 shows you some elementary circuits you can build to demonstrate the principles of electronics and observe specific electronic components functioning as advertised. Turn to this chapter if you want to reinforce your theoretical knowledge of electronics or gain experience in building simple circuits. When you're ready for more complex circuits, explore Chapter 15. There you find several projects that you can have fun building and exploring. You may even decide to put one or two of them to good use in your home or office.

Part IV: The Part of Tens

As you might expect, Part IV is where you can find additional electronics-related information, laid out in top-ten list format. Chapter 16 offers pointers to help you expand your electronics horizons. There, you can find information on all-inclusive project kits and circuit simulation software, suggestions for additional testing tools, and tips on how to find deals on electronics supplies. When you're ready to shop for all things electronic, turn to Chapter 17 for a list of top-notch electronics suppliers in the United States and abroad.

Icons Used in This Book

Because we can't place dozens of sticky-note flags in each and every *Electronics For Dummies* book, we use graphical icons to draw your attention to critical information that stands out in one way or another.

Tips alert you to information that can truly save you time, headaches, or money (or all three!). You'll find that if you use our tips, your electronics experience will be that much more enjoyable.

When you tinker with electronics, you're bound to encounter situations that call for extreme caution. Enter the Warning icon, a not-so-gentle reminder to take extra precautions to avoid personal injury or prevent damage to your tools, components, circuits — or your pocketbook.

This icon reminds you of important ideas or facts that you should keep in mind while exploring the fascinating world of electronics. Occasionally, we use this icon to note where in the book an important concept is originally introduced, so you can flip back to more detailed information for a refresher, if you need one.

Even though this entire book is about technical stuff, we flag certain topics to alert you to deeper technical information that might require a little more brain power to digest. Of course, if you choose to skip over this information, that's okay — you can still follow along just fine. Think of this information as extra material — a diversion off the main path, if you will — like extra credit questions on a math test.

Part I
Understanding the Fundamentals of Electronics

The 5th Wave   By Rich Tennant

"So I guess you forgot to tell me to strip out the components before drilling for blowholes."

In this part . . .

Do you have a burning desire to understand what makes electronic devices tick? Have you been curious to know how speakers speak, motors move, and computers compute? Well, then, you've come to the right place!

In the chapters ahead, we explain exactly what electronics is, what it can (and does) do for you, and how all sorts of electronic devices work. Don't worry: We don't bore you with long essays involving physics and mathematics — even though we could. We use analogies and down-to-earth examples involving water, marbles, and desserts to make it easy — fun, even — to understand. And, while you're enjoying yourself, you gain a fairly deep understanding of how electronic components work and combine forces to make amazing things happen.

Chapter 1

What Is Electronics and What Can It Do for You?

*I*f you're like most people, you probably have some idea about what electronics is. You've been up close and personal with lots of so-called "consumer electronics" devices, such as iPods, stereo equipment, personal computers, digital cameras, and televisions, but to you, they may seem like mysteriously magical boxes with buttons that respond to your every desire.

You know that underneath each sleek exterior lies an amazing assortment of tiny components connected together in just the right way to make something happen. And now you want to understand how.

In this chapter, you find out that electrons moving in harmony constitute electric current — and that controlling electric current is the basis of electronics. You take a look at what electric current really is and what you need to keep the juice flowing. You also get an overview of some of the things you can do with electronics.

Just What Is Electronics?

When you turn on a light in your home, you're connecting a source of electrical energy (usually supplied by your power company) to a light bulb in a complete path, known as an *electrical circuit*. If you add a dimmer or a timer to the light bulb circuit, you can *control* the operation of the light bulb in a more interesting way than simply switching it on and off.

Electrical systems, such as the circuits in your house, use pure, unadulterated electric current to power things like light bulbs. *Electronic systems* take this a step further: They *control* the current, changing its fluctuations, direction, and timing in various ways in order to accomplish a variety of functions, from dimming a light bulb to communicating with satellites (and lots of other things). (See Figure 1-1.) It is this control that distinguishes electronic systems from electrical systems.

To understand how electronics involves the control of electric current, first you need a good working sense of what electric current really is and how it powers things like light bulbs.

Figure 1-1:
The dimmer electronics in this circuit control the flow of electric current to the light bulb.

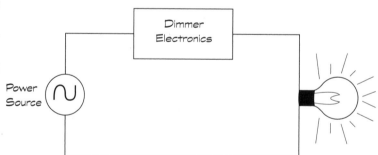

What is electricity?

The simple truth about electricity is that it is not so simple. The term "electricity" is ambiguous, often contradictory, and can lead to great confusion, even among scientists and teachers.

Generally speaking, "electricity" has do with how certain types of particles found in nature interact with each other when a bunch of them are hanging around in the same general area.

Rather than talk about electricity, you're better off using other, more precise, terminology to describe all things electric. Here are some of them:

✔ **Electric charge:** A fundamental (that means don't question it) property of certain particles that describes how they interact with each other. There are two types: positive and negative. Particles of the same type (positive or negative) repel each other, while particles of the opposite type attract each other.

✔ **Electrical energy:** A form of energy caused by the behavior of electrically charged particles. This is what you pay your electric company to supply.

✔ **Electric current:** The flow of electrically charged particles. This is probably the connotation of electricity you are most familiar with, and the one we focus on in this chapter.

So, if you're just bantering around the water cooler, it's okay to use the word electricity to describe the stuff that powers your favorite gaming system, but if you throw that word around carelessly among learned physics types, you might just repel them.

Checking Out Electric Current

Electric current, sometimes known as electricity (see the sidebar "What is electricity?"), is the flow of teeny tiny electrically charged particles called *electrons.* So where exactly do you find electrons, and how do they move around? You'll find the answers by taking a peek inside the atom.

Getting a charge out of electrons

Atoms are the basic building blocks of everything in the universe, whether natural or manmade. They're so tiny, you'd find millions of them in a single speck of dust, so you can imagine how many there are in your average sumo wrestler. Electrons can be found in every single atom in the universe, living outside the atom's center, or *nucleus.* All electrons carry a negative electric charge and are attracted to other tiny particles called *protons,* which carry a positive electric charge and exist inside the nucleus.

Electric charge is a property of certain particles, such as electrons, protons, and quarks (yes, quarks), that describes how they interact with each other. There are two different flavors of electric charge, somewhat arbitrarily named "positive" and "negative" (okay, you really could call them "Moe" and "Larry" or "north" and "south" instead, but those names are already taken). In general, particles carrying the same type of charge repel each other, whereas particles carrying different charges attract each other. That's why electrons and protons find each other so attractive.

Under normal circumstances, there are an equal number of protons and electrons in each atom, and the atom is said to be *electrically neutral.* The attractive force between the protons and electrons acts like invisible glue, holding the atomic particles together, in much the same way that the gravitational force of the Earth keeps the moon within sight. The electrons closest to the nucleus are held to the atom with a stronger force than the electrons farther from the nucleus; some atoms hold on to their outer electrons with a vengeance while others are a bit more lax.

Mobilizing electrons in conductors

Materials (such as air or plastic) that like to keep their electrons close to home are called *insulators.* Materials, such as copper, aluminum, and other metals, that contain loosely bound outer electrons are called *conductors.*

In metals, the outer electrons are bound so loosely, many of them break free and wander around among the metal atoms. These "free" electrons are like sheep grazing on a hillside: They drift around aimlessly but don't move very far or in any particular direction. But if you give these free electrons a bit of a push in one direction, they will gladly move in the direction of the push. *Electric current* (often called electricity) is the movement *en masse* of electrons through a conductor when an external force (or push) is applied.

Figure 1-2:
Electron flow through a conductor is analogous to a bucket brigade.

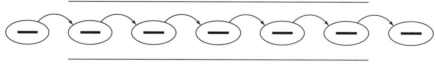

This flow of electric current appears to happen instantaneously. That's because each free electron — from one end of a conductor to the other — begins to move more or less immediately.

Think of a bucket brigade: You've got a line of people, each holding a bucket of water, with a person at one end filling an empty bucket with water, and a person at the other end dumping a full bucket out. On command, each person passes his bucket to his neighbor on the left, and accepts a bucket from his neighbor on the right, as in a bucket brigade. Although each bucket moves just a short distance (from one person to the next), it appears as if a bucket of water is being transported from one end of the line to the other. Likewise, with electric current, as each electron displaces the one in front of it along a conductive path, it appears as if the electrons are moving nearly instantaneously from one end of the conductor to the other. (See Figure 1-2.)

Electric current is a realm of tiny things that sometimes interact in huge quantities, so it needs its own units of measurement. A *coulomb,* for example, is defined as the charge carried by 6.24×10^{18} (that's 624 followed by 16 zeros) electrons. If a coulomb of charge moves past a point within a second, we say that the strength of the electric current is *one ampere,* or one amp (abbreviated as 1 A). That's a whole lot of electrons at once, much more than are typically found in electronic systems. There you're more likely to see current measured in *milliamps* (mA). A milliamp is one one-thousandth of an amp.

Giving electrons a nudge

Electric current is the flow of negatively charged electrons through a conductor when a force is applied. But just what is the force that provokes the electrons to move in harmony? What commands the electronic bucket brigade?

The force that pushes electrons along is known as *voltage,* and it is measured in units called *volts* (abbreviated V). Apply enough voltage to a conductor, and the free electrons within it will move together in the same direction, like sheep begin herded into a pen — only much faster.

Think of voltage as electric pressure. In much the way water pressure pushes water through pipes and valves, voltage pushes electrons through conductors. The higher the pressure, the stronger the push — so the higher the voltage, the stronger the electric current that flows through a conductor.

You may also hear the terms *potential difference, voltage potential, potential drop,* or *voltage drop* used to describe voltage. Try not to let these different terms confuse you. There's more about this in Chapter 2.

Experiencing electricity

You can personally experience the flow of electrons by shuffling your feet across a carpet on a dry day and touching a doorknob; that zap you feel (and the spark you may see) is the result of electrically charged particles jumping from your fingertip to the doorknob, a form of electricity known as *static electricity*. Static electricity is an accumulation of electrically charged particles that remain static (unmoving) until drawn to a bunch of oppositely charged particles.

Lightning is another example of static electricity (but not one you want to experience personally), with charged particles traveling from one cloud to another or from a cloud to the ground. When charged particles move around, they release energy (hence the zaps and the sparks).

If you can get enough charged particles to move around, and you can harness the energy they release, you can use that energy to power light bulbs and other things.

Harnessing Electrical Energy to Do Work

Ben Franklin was one of the first people to observe and experiment with electricity, and he came up with many of the terms and concepts (for instance, *current*) we know and love today. Contrary to popular belief, Franklin didn't actually *hold* the key at the end of his kite string during that storm in 1752. (If he had, he wouldn't have been around for the American Revolution.) He may have performed that experiment, but not by holding the key.

Franklin knew that electricity was both dangerous and powerful, and his work got people wondering whether there was a way to use the power of electricity for practical applications. Scientists like Michael Faraday, Thomas Edison, and others took Franklin's work a bit further and figured out ways to harness electrical energy and put it to good use.

As you begin to get excited about harnessing electrical energy, take note of the scary-looking Warning icon to the left, and remember that over 250 years ago, Ben Franklin knew enough to be careful around the electrical forces of nature. And so should you. Even tiny amounts of electric current can be quite dangerous — even fatal — if the circumstances are right (or wrong). In Chapter 9, we explain more about the harm current can inflict and the precautions you can (and must) take to stay safe when working with electronics. But for now, consider this a warning!

In this section, we explore how electrons transport energy — and how that energy can be applied to make things work.

Tapping into electrical energy

As electrons travel through a conductor, they transport energy from one end of the conductor to the other. Because like charges repel, each electron exerts a non-contact repulsive force on the electron next to it, pushing that electron along through the conductor. As a result, electrical energy is propagated through the conductor.

If you can transport that energy to an object that allows work to be done on it, such as a light bulb, a motor, or a loudspeaker, you can put that energy to good use. The electrical energy carried by the electrons is absorbed by the object and transformed into another form of energy, such as light, heat, or mechanical energy. That's how you make the filament glow, rotate the motor shaft, or cause the diaphragm of the speaker to vibrate.

Because you can't see — and you don't necessarily want to touch — gobs of flowing electrons, try thinking about water to help make sense out of harnessing electrical energy. A single drop of water can't do much to help (or hurt) anyone, but get a whole group of water drops to work in unison, funnel them through a conduit, direct the flow of water toward an object (for example, a waterwheel), and you can put the resulting water energy to good use. Just as millions of drops of water moving in the same direction constitute a current, millions of electrons moving in the same direction make an electric current. In fact, Benjamin Franklin came up with the idea that electricity acts like a fluid and has similar properties, such as current and pressure (but he probably would have cautioned you against drinking it).

But where does the original energy — the thing that starts the electrons moving in the first place — come from? It comes from a source of electrical energy, such as a battery (we discuss electrical energy sources in Chapter 2).

Making sure electrons arrive at their destination

Electric current doesn't flow just anywhere. (If it did, you'd be getting shocked all the time.) Electrons only flow if you provide a closed conductive path, or *circuit,* for them to move through, and initiate the flow with a battery or other source of electrical energy. Copper and other conductors are commonly formed into wire to provide a path for the flow of free electrons, so you can direct electrical energy to a light bulb or other object that will use it. Just as with pipes and water, the wider the wire, the more freely the electrons flow.

Working electrons deliver power

To electrons delivering energy to a light bulb or other device, the word "work" has real physical meaning. *Work* is a measure of the energy consumed by the device over some time when a force (voltage) is applied to a bunch of electrons in the device. The more electrons you push, and the harder you push them, the more electrical energy is available and the more work can be done (for instance, the brighter the light, or the faster the motor rotation). The total energy consumed in doing work over some period of time is known as *power* and is measured in *watts*. Power is calculated by multiplying the force (voltage) by the strength of the electron flow (current):

Power = voltage × current

Power calculations are really important in electronics, because they help you understand just how much energy electronic parts are willing (and able) to handle without complaining. If you energize too many electrons in the same electronic part, you'll generate a lot of heat energy and you might fry that part. Many electronic parts come with maximum power ratings so you can avoid getting into a heated situation. We remind you about this in later chapters when we discuss specific components and their power ratings.

If there's a break in the path (an *open circuit*), electrons stop flowing — and the metal atoms in the wire quickly settle down to a peaceful, electrically neutral existence. Picture a gallon of water flowing through an open pipe. The water will flow for a short time, but then stop when all the water exits the pipe. If you pump water through a closed pipe system, the water will continue to flow as long as you keep forcing it to move. To keep the electrons flowing, you need to connect everything together in one big happy *electrical circuit*. As shown in Figure 1-3, every circuit needs at least three basic things to ensure that electrons get energized and deliver their energy to something that needs work done:

- ✔ **A *source* of electrical energy:** The source provides the force that nudges the electrons through the circuit. You may also hear the terms *electrical source, power source, voltage source,* and *energy source* used to describe a source of electrical energy. We discuss sources of electrical energy in Chapter 2.

- ✔ **A *load:*** The load is something that absorbs electrical energy in a circuit (for instance, a light bulb or a speaker). Think of the load as the destination for the electrical energy.

- ✔ **A *path:*** A conductive path provides a conduit for electrons to flow between the source and the load.

An electric current starts with a "push" from the source and flows through the wire path to the load, where electrical energy makes something happen — emitting light, for instance.

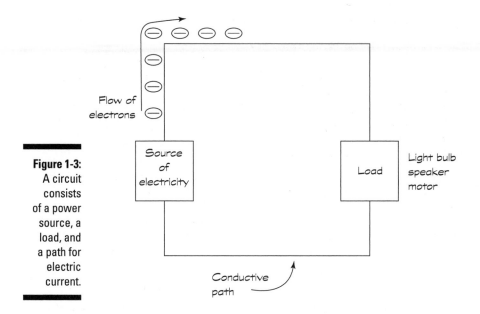

Oh, the Things Electrons Can Do (Once You Put Their Minds to It)!

Imagine applying an electric current to a pair of speakers without using anything to control or "shape" the current. What would you hear? Guaranteed it wouldn't be music! By using the proper combination of electronics assembled in just the right way, you can control the way each speaker diaphragm vibrates, producing recognizable sounds such as speech or music (well, certain music anyway). There's so much more you can do with electric current once you know how to control the flow of electrons.

Electronics is all about using specialized devices known as *electronic components* (for example, resistors, capacitors, inductors, and transistors, which we discuss in Chapters 3, 4, 5, and 6, respectively) to control current (also known as the flow of electrons) in such a way that a specific function is performed.

Simple electronic devices use a few components to control current flow. The dimmer switch that controls current flowing into a light bulb is one such example. But most electronic systems are a lot more complicated than that; they connect lots of individual components together in one or more circuits to achieve their ultimate goal. The nice thing is that you once you understand how a few individual electronic components work and how to apply some basic principles, you can begin to understand and build interesting electronic circuits.

This section provides just a sampling of the sorts of things you can do by controlling electrons with electronic circuits.

Creating good vibrations

Electronic components in your iPod, car stereo, and other audio systems convert electrical energy into sound energy. In each case, the system's speakers are the load, or destination, for electrical energy, and the job of the electronic components within the system is to "shape" the current flowing to the speakers so that the diaphragm within each speaker moves in such a way as to reproduce the original sound.

Seeing is believing

In visual systems, electronic components control the timing and intensity of light emissions. Many remote-control devices, such as the one wedged in your La-Z-Boy recliner, emit infrared light when you press a button, and the specific pattern of the emitted light acts as a sort of code to the device you are controlling, telling it what to do.

The inside surface of the tube in a cathode-ray tube (CRT) TV set (are there any still around?) is coated with phosphors that glow when struck by electron beams within the tube. The electronic circuits within the TV set control the direction and intensity of the electron beams, thus controlling the pattern painted across the TV screen — which is the image you see. Enlightening, isn't it?

Sensing and alarming

Electronics can also be used to make something happen in response to a specific level of light, heat, sound, or motion. Electronic *sensors* generate or change an electrical current in response to a stimulus. Microphones, motion detectors, temperature sensors, and light sensors can be used to trigger other electronic components to perform some action, such as activating an automatic door opener or sounding an alarm.

Controlling motion

A common use of electronics is to control the on/off activity and speed of motors. By attaching various objects — for instance, wheels, airplane flaps, or your good-for-nothing brother-in-law — to motors, you can use electronics to control their motion. Such electronics can be found in robotic systems, aircraft, spacecraft, elevators, and lots of other places.

Solving problems (a.k.a. computing)

In much the same way that the ancients (those living long ago, not your great-grandparents) used the abacus to perform arithmetic operations, so you use electronic calculators and computers to perform computations. With the abacus, beads were used to represent numbers, and calculations were performed by manipulating those beads. In computing systems, patterns of stored electrical energy are used to represent numbers, letters, and other information, and computations are performed by manipulating those patterns using electronic components. (Of course, the worker-bee electrons inside have no idea they are crunching numbers!) If you have your decoder ring handy, you can translate the resulting pattern into an actual number (or you can just let the display electronics do that for you).

Communicating

Electronic circuits in your cellphone work together to convert the sound of your voice into an electrical pattern, manipulate the pattern (to compress and encode it for transmission), convert it into a radio signal, and send it out through the air to a communication tower. Other electronic circuits in your handset detect incoming messages from the tower, decode the messages, and convert an electrical pattern within the message into the sound of your friend's voice (via a speaker).

Data-communication systems, which you use every time you shop online, use electronics to convert your materialistic desires into shopping orders — and (usually) extract money from your bank account.

Chapter 2

Manipulating Electricity to Make Something Happen

*E*lectronics is all about controlling the flow of electrons (electric current) through conductors in a complete path (circuit) so that the electrical energy delivered to a load (such as a light bulb, motor, or speaker) is "shaped" in just the right way. By manipulating the flow of electrons, electronic components enable you to do some amazing things with electricity, such as vary the sound produced by speakers, change the direction and speed of motors, and control the intensity and timing of lights, among many other things. In other words, electronics doesn't make electricity — it makes electricity better.

In this chapter, you discover how to get electrons flowing through a circuit and why conventional current can be thought of as electrons moving in reverse. You also explore the depths of a simple electronic circuit, and look at different ways to connect electronic components so you can begin to shape and direct current the way you see fit in your own circuits. Finally, you'll take a look at how two familiar electronic devices — your radio and your TV set — manipulate electric current to make your life more entertaining.

Supplying Electrical Energy

If you take a copper wire and arrange it in a circle by twisting the ends together, do you think the free electrons will flow? Well, the electrons might dance around a bit, because they're so easy to move, but unless there's a force pulling them one way or another, you won't get current to flow.

Think about the motion of water that is just sitting in a closed pipe: The water may bounce up and down a bit, but it's not going to go whooshing through the pipe on its own. You need to introduce a force, a pressure differential, in order to deliver the energy needed to get a current flowing through the pipe.

A circuit needs a *source of electricity* (really, electrical energy) to get the electrons flowing. Batteries and solar cells are common sources; the electrical energy available at your wall outlets may come from one of many different sources supplied by your power company. But what exactly is a source of electricity? How do you "conjure up" electrical energy?

All sources of electricity work by converting another form of energy (for instance, mechanical, chemical, heat, light) into electrical energy. Exactly how electrical energy is generated by your favorite source turns out to be important, because different sources produce different types of electric current. The two different types are

- ✔ **Direct current (DC):** A steady flow of electrons in one direction, with very little variation in the strength of the current. Cells (commonly known as batteries) produce DC and most electronic circuits use DC.

- ✔ **Alternating current (AC):** A fluctuating flow of electrons that changes direction periodically. Power companies supply AC to your electrical outlets.

Getting direct current from a battery

A battery converts chemical energy into electrical energy through a process called an *electrochemical reaction.* When two different metals are immersed in a certain type of chemical, the metal atoms react with the chemical atoms to produce charged particles. Negative charges build up on one of the metal plates, while positive charges build up on the other metal plate. The difference in charge across the two metal terminals (a *terminal* is just a piece of metal to which you can hook up wires) creates a voltage. That voltage is the force that electrons need to push them around a circuit.

To use a battery in a circuit, you connect one side of your load — for instance, a light bulb — to the negative terminal (known as the *anode*) and the other side of your load to the positive terminal (known as the *cathode*). You've created a path that allows the charges to move, and electrons flow from the anode, through the circuit, to the cathode. As they pass through the wire filament of the light bulb, some of the electrical energy supplied by the battery is converted to heat, causing the filament to glow.

Because the electrons move in only one direction (from the anode, through the circuit, to the cathode), the electric current generated by a battery is DC. (See Figure 2-1.) A battery continues to generate current until all the chemicals inside it have been used up by the electrochemical process. The AAA-, AA-, C-, and D-size batteries you can buy almost anywhere each generate about 1.5 volts — regardless of size. The difference in size among those batteries has to do with how much current can be drawn from them. The larger the battery, the more current can be drawn, and the longer it will last. Larger batteries can handle heavier loads, which is just a way of saying they can produce more power (remember, power = voltage × current), so they can do more work.

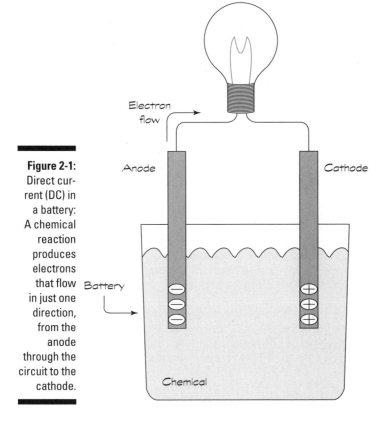

Figure 2-1:
Direct current (DC) in a battery: A chemical reaction produces electrons that flow in just one direction, from the anode through the circuit to the cathode.

Technically speaking, an individual "battery" isn't really a battery (that is, a group of units working together); it's a *cell* (one of those units). If you connect several cells together, as you often do in many flashlights and children's toys, *then* you've created a battery. The battery in your car is made up of six cells, each generating 1.5 volts, connected together to produce 12 volts total. We discuss various types of cells — and how to connect them to create higher voltages — in Chapter 8.

This is the symbol commonly used to represent a battery in a circuit diagram. The plus sign signifies the cathode; the minus sign signifies the anode. Usually the battery's voltage is shown alongside the symbol.

Using alternating current from a power plant

When you plug a light into an electrical outlet in your home, you're using electrical energy that originated at a generating plant. Generating plants process natural resources — such as water, coal, oil, natural gas, or uranium — through several steps to produce electrical energy. That's why electrical energy is said to be a *secondary* energy source: It's generated through the conversion of a primary energy source.

Many plants use the heat energy generated by nuclear reactions or the burning of fossil fuels to transform water into steam. Then the steam exerts pressure on the fins of a turbine, causing it to turn. Hydroelectric power plants located at dams use water pressure, and windmills use wind energy to rotate turbines. Power-plant turbines are connected to electromechanical generators, which convert mechanical energy (in this case, the motion of the turbine) into electrical energy. A generator contains a coil of wire inside a huge permanent magnet. As the turbine rotates, it turns the coil of wire, and — presto! — electrical current is *induced* in the wire. That's just a technical way of saying something is causing electrons to flow, without any direct contact with the wire.

Electron flow can be induced by moving a wire near a magnet, or moving a magnet near a wire. This is called *electromagnetic induction*, and it has to do with the close relationship between magnetism and electricity. We mention this again in Chapter 5 when we discuss inductors.

As the coil rotates inside the magnet, the magnet first causes the electrons to flow in one direction, but when the coil has rotated 180 degrees, the magnet pulls the electrons in the other direction. You might say that the electrons "pull a 180" and switch direction! This rotation creates *alternating current* (AC).

Doing the (sine) wave

Because alternating current is constantly changing, you can't describe its strength with a single number, as you can with DC. A common way to discuss its variations is to look at a *waveform*, or the pattern of the current over time. The AC current waveform shows the fluctuations of current, with "positive current" representing electron flow in one direction, and "negative current" representing electron flow in the other direction. The *instantaneous current* is the strength of the current at a single point in time, and *peak current* is the magnitude (absolute value) of the current at its highest and lowest points. Because you can use the mathematical sine function to calculate the current for a given time, AC waveforms are often referred to as *sine waves.* (If you think you smell trigonometry here, you're right, but don't worry — you don't need to wipe the dust off your high-school math books! We just want you to be aware of the term "sine" as it's used in electronics.)

Waveforms are also used to describe a fluctuating voltage, commonly called *AC voltage.* The *peak voltage* (symbolized by V_p) is the magnitude of the highest voltage. You may hear the term *peak-to-peak voltage* (symbolized by V_{pp}), which is a measure of the difference between the highest and lowest voltage on the waveform, or twice the peak value. Another term thrown around is *rms voltage* (symbolized by V_{rms}), which is short for *root-mean-square voltage,* and it's used in power calculations as a way to compare the effects of AC power with DC power. There's a mathematical formula to calculate this value, but it ends up being 0.7071 times the peak value of the voltage.

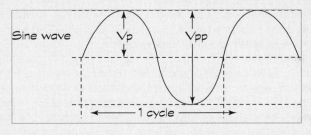

In power plants in the United States, the coil makes 60 complete rotations each second, so the electron flow changes direction 120 times each second. When the change in electron flow makes a complete loop (which it does 60 times a second), it's called a *cycle.* The number of cycles per second in alternating current is known as *frequency* and is measured in units called *hertz,* abbreviated Hz. The United States and Canada generate AC at 60 hertz, while the most European countries use 50 hertz as a standard. You can be pretty sure that any country you visit uses either 50 Hz or 60 Hz line current.

This symbol is used in circuit diagrams for an AC voltage source. AC is usually generated at 13,800 volts and then *stepped up* (transformed to higher voltages) for transmission across long distances. After it reaches its destination, it's *stepped down* (reduced in voltage) to 240 volts or 120 volts for distribution to homes and businesses. So the electricity supplied by your average wall outlet is said to be 120 volts AC (or 120 VAC), which just means it's alternating current at 120 volts.

Heaters, lamps, hair dryers, and electric razors are among the electrical devices that use 120 volts AC directly; clothes dryers, which require more power, use 240 volts AC directly from a special wall outlet. If your hair dryer uses 60 Hz power, and you're visiting a country that uses 50 Hz power, you'll need a *power converter* to get the hertz you need from your host country.

Many electronic devices (for instance, laptop computers) require a steady DC supply, so if you're using AC to supply an electronic device or circuit, you'll need to convert AC to DC. *Regulated power supplies,* also known as *AC-to-DC adapters,* or *AC adapters,* don't actually *supply* power: They convert AC to DC and are commonly included with electronic devices when purchased. Think of your cellphone charger; this little device essentially converts AC power into DC power that the battery in your cellphone uses to charge itself back up.

Transforming light into electricity

Solar cells, also known as *photovoltaic cells,* produce a small voltage when you shine light on them. They are made from *semiconductors,* which are materials that are somewhere between conductors and insulators in terms of their willingness to give up their electrons. (We discuss semiconductors in detail in Chapter 6.) The amount of voltage produced by a solar cell is fairly constant, no matter how much light you shine on it — but the *strength* of the current you can draw depends on the intensity of the light: The brighter the light, the higher the strength of the available current.

Solar cells have wires attached to two terminals for conducting electrons through circuits, so you can power your calculator or the garden lights that frame your walkway. You may have seen arrays of solar cells used to power emergency road signs, call boxes, or lights in parking lots, but you probably haven't seen the large solar-cell arrays used to power satellites (not from close up, anyway). Solar panels are becoming increasingly popular for supplying electrical power to homes and businesses as a way to reduce utility costs. If you scour the Internet, you'll find lots of information on how you can make your own solar panels — for just a couple of hundred dollars and a willingness to try. You can read more about this in *Solar Power Your Home For Dummies* by Rik DeGunther (Wiley Publishing, Inc.).

Understanding Directions: Real Electron Flow versus Conventional Current Flow

Early experimenters believed that electric current was the flow of positive charges, so they described electric current as the flow of a positive charge from a positive terminal to a negative terminal. Much later, experimenters discovered electrons and determined that they flow from a negative terminal to a positive terminal. That original convention is still with us today — so the standard is to depict the direction of electric current in diagrams with an arrow that points opposite to the direction of actual electron flow.

Conventional current is the flow of a positive charge from positive to negative voltage and is just the reverse of real electron flow. (See Figure 2-2.) All descriptions of electronic circuits use conventional current, so if you see an arrow depicting current flow in a circuit diagram, you know it is showing the direction of conventional current flow. In electronics, the symbol *I* represents conventional current, measured in amperes (or amps, abbreviated A). You're more likely to encounter *milliamps* (mA) in circuits you build at home. A milliamp is one one-thousandth of an amp.

Figure 2-2:
Conventional current, I, flows from the positive side of a power source to the negative side; real electrons flow from the negative side to the positive side.

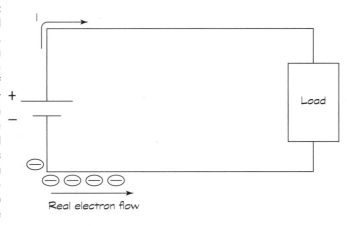

In AC circuits, current is constantly reversing direction. So how do you show current flow in a circuit diagram? Which way should the arrow point? The answer is that it doesn't matter. You arbitrarily pick a direction for the current flow (known as the *reference direction*), and you label that current *I*. The value of I fluctuates up and down as the current alternates. If the value of I is negative, that just means that the (conventional) current is flowing in the direction opposite to the way the arrow is pointing.

Examining a Simple Light-Bulb Circuit

The diagram in Figure 2-3 depicts a battery-operated circuit that lights a bulb, much like what you might find in a flashlight. What you see in the figure is a circuit diagram, or *schematic,* that shows all the components of the circuit and how they are connected. (We discuss schematics in detail in Chapter 10.)

The battery is supplying 1.5 volts DC to the circuit. That just means it supplies a steady 1.5 volts. The plus sign near the battery symbol indicates the positive terminal of the battery, from which current flows (conventional current, of course). The negative sign near the battery symbol indicates the negative terminal of the battery, to which the current flows after it makes its way around the circuit. The arrow in the circuit indicates the reference direction of current flow, and because it's pointing away from the positive terminal of the battery in a DC circuit, you should expect the value of the current to be positive all the time.

Figure 2-3:
Current from a battery flows through the circuit, delivering electrical energy to the light bulb as voltage is applied to the bulb.

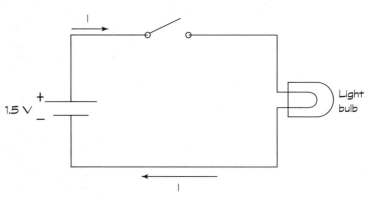

The lines in the circuit diagram show how the circuit components are connected, using wire or other connectors. We discuss various kinds of wire and connectors in Chapter 8. Switches and other circuit components are usually made with *leads* — protruding wires connected to the innards of the component that provide the means to connect the component to other circuit elements.

Next to the battery is a switch. This simply opens and closes the circuit, allowing current to flow out of the battery or stopping it dead in its tracks. If the switch is closed, current flows out of the battery, through the light bulb, where electrical energy is dissipated as light and heat, and back into the negative terminal of the battery. If the switch is open, current will not flow at all through this *open circuit.*

The battery is *supplying* electrical energy and the light bulb is *using* electrical energy (actually, it's converting electrical energy into heat energy). There's a give-and-take relationship between the two: Voltage is the push the battery gives to get current moving, and energy from that push is absorbed when current moves through the light bulb. As current flows through the bulb, voltage drops across the bulb. Think of it as if the bulb is using up energy supplied by the force (voltage) that pushes the current through it.

When you "drop voltage" across a light bulb or other component, the voltage is higher where the current enters the component than is the case where the current exits the component. Voltage is actually a relative measurement, because it's the force that results from a difference in charge from one point to another. The voltage supplied by a battery represents the difference in charge from the positive terminal to the negative terminal, and that difference in charge has the potential to move current through a circuit; the circuit, in turn, absorbs the energy generated by that force as the current flows, which drops the voltage. This is why voltage is sometimes called *voltage drop, potential difference,* or *potential drop.*

When you see references to the voltage *at a single point* in a circuit, you should know that it is always with respect to the voltage at another point in the circuit — usually the *reference ground* (often simply called *ground*), the point in the circuit that is (arbitrarily) said to be at 0 volts. Often, the negative terminal of a battery is used as the reference ground, and all voltages throughout the circuit are measured with respect to that reference point.

An analogy that may help you understand voltage measurement is distance measurement. If someone were to ask you, "What's your distance?" you'd probably say, "Distance from what?" Similarly, if you're asked, "What's the voltage at the point in the circuit where the current enters the light bulb?" you should ask, "With respect to what point in the circuit?" On the other hand, you may say, "I'm five miles from home," and you've stated your distance from a reference point (home). So if you say, "The voltage where the current enters the bulb is 1.5 volts with respect to ground," that makes perfect sense.

If you start at the negative terminal of the battery in the simple light-bulb circuit, and travel all the way around the circuit measuring voltages, you'll see that the voltage at the positive terminal of the battery is 1.5 volts, and all of those 1.5 volts are dropped across the light bulb. (In reality, the switch eats up a teeny tiny bit of voltage, because even the best conductors use up some energy, but it's negligible compared to the voltage drop across the light bulb.)

The important thing to notice here is that as you travel around a DC circuit, you "gain" voltage going from the negative terminal of the battery to the positive terminal (that's known as a *voltage rise*), and you "lose" or drop voltage as you continue in the same direction across circuit components. By the time you get back to the negative terminal of the battery, all the battery voltage has been dropped and you're back to 0 volts. With all circuits (whether AC or DC), if you start at *any* point in the circuit, and add up the voltage rises and drops going around the circuit, you will end up back where you started. The net sum of the voltage rises and drops in voltage around a circuit is zero.(This is known as *Kirchoff's Voltage Law.*)

Standing your ground

In electronics, the word "ground" can have two different meanings. *Earth ground* means pretty much what it says: It's a direct connection to the ground — real ground, the stuff of the planet. The screw in the center of a standard two-prong AC outlet, as well as the third prong in a three-prong outlet, is connected to earth ground. Behind each wall socket is a wire that runs through your house or office and eventually connects to a metal post that makes good contact with the ground. This arrangement provides extra protection for circuits that use large amounts of current; shipping dangerous current directly into the earth gives it a safe place to go after it's done its work — or gives it something to do besides destroy things. Such was the case when Ben Franklin's lightning rod provided a direct path for dangerous lightning to hit the ground — instead of a house or person. In circuits that handle large currents, some point in the circuit is usually connected to a pipe or other metal object that's connected to earth ground.

The term *floating ground* refers to a circuit that is not connected to earth ground — which may be dangerous. You'd be wise to stay away from such a circuit until it is safely grounded (or "earthed," as folks say in the U.K.)!

The other type of ground is called a *common ground,* or simply *common.* It isn't a physical ground; rather, it is a just a reference point within a circuit for voltage measurements. Certain types of circuits, particularly the circuits commonly used in computers, label the negative terminal of a DC power supply the common ground, and connect the positive terminal of another DC power supply to the same point. That way, the circuit is said to have both positive and negative power supplies. The two physical power supplies may be identical, but the way you connect them in a circuit and the point you choose for the zero voltage reference determine whether a supply voltage is positive or negative. It's all relative!

Keep in mind that these voltage drops have real physical meaning. The electrical energy supplied by the battery is absorbed by the light bulb. The battery will keep supplying electrical energy and shipping out current, and the light bulb will keep absorbing that energy, until the battery "dies" — runs out of energy. That happens when all the chemicals inside the battery have been consumed in the chemical reactions that produced the positive and negative charges. In effect, all the chemical energy supplied by the battery has been converted into electrical energy — and absorbed by the circuit.

One of the fundamental laws of physics is that energy cannot be created or destroyed; it can only change form. You really see this first-hand with the simple battery-driven light bulb circuit: Chemical energy is converted to electrical energy, which is converted to heat and light energy, which — well, you get the idea.

You can measure the voltage drop across the light bulb using a *voltmeter* (which we discuss in detail in Chapter 12). And if you multiply the *voltage across the bulb* (that's a common way of saying "voltage drop") by the strength of the current running through the bulb, you get the power dissipated in the bulb (power = voltage × current) in watts.

Controlling Electrical Current with Basic Components

If you were to build the simple light-bulb circuit discussed in the previous section, and you didn't have a 1.5 V battery available, you might think it would be okay to use the 9 V battery you found in the kitchen drawer. After all, 9 V is more than 1.5 V, so the battery should provide enough energy to light the bulb. As it turns out, if you use the 9 V battery, your little circuit will draw a lot more current — and you may overload your light bulb. If too many electrons are allowed to flow through a filament, the electrical energy dissipated in the filament will create so much heat that the bulb will burst.

What you can do is insert a little electronic device called a *resistor* between the battery and the light bulb. Resistors restrict the flow of current through a circuit, and are commonly used to protect other circuit elements (such as light bulbs) from receiving more electrons than they can handle. The resistor is just one electronic component that controls the flow of current in a circuit, but there are many more.

Ways to control current

Controlling electrical current is similar in many ways to controlling H_2O current. How many different ways can you control the flow of water using various plumbing devices and other components? Some of the things you can do are restrict the flow, cut off the flow completely, adjust the pressure, allow water to flow in one direction only, and store water. (This water analogy may help but it isn't 100% valid; you don't need a closed system for water to flow — and you *do* need a closed system to make electric current flow.)

Empowering you to make the right choices

Light bulbs and other electronic components have maximum power ratings for good reason. Send too much current through them, and they overheat and burn or melt. Remember that power is the product of voltage and current, so once you understand how to figure out voltage drops and the amount of current passing through these components, it will be completely within your power to estimate the *power rating* (that is, how many watts the part can handle before blowing up in your face) you need for the components you select for your circuits.

There are many, many electronic components that help control the electrical energy in circuits. (See Figure 2-4.) Among the most popular components are *resistors*, which restrict current flow, and *capacitors*, which store electrical energy. (We discuss resistors at length in Chapter 3 and capacitors in Chapter 4.) *Inductors* and *transformers* are devices that store electrical energy in magnetic fields. (You can get the details about them in Chapter 5.) *Diodes* are used to restrict current flow in one direction, much like valves, while *transistors* are versatile components that can be used to switch circuits on and off, or amplify current. (We cover diodes and transistors in Chapter 6.)

Active versus passive components

You may see the terms *active components* and *passive components* used as category headings for types of electronic components. *Active components* are parts that provide gain to current (that is, boost it) or direct it; examples are transistors and diodes. (These may also be categorized as *semiconductors,* which refers to the type of material they're made from.) *Passive components* provide neither gain (amplification) nor direction — though they can slow current or store electrical energy — so resistors, capacitors, inductors, and transformers are all passive components. (Step-up transformers increase voltage while decreasing current.) A circuit that contains only passive components is called a *passive circuit;* one that contains *at least one* active component is an *active circuit.*

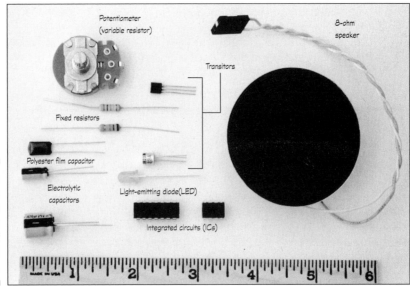

Figure 2-4: Electronic components come in a variety of shapes and packages.

Making Connections: Series and Parallel

Just as you can build structures of all shapes and sizes by connecting LEGOs or K'NEX pieces in various ways, so you can build many different kinds of circuits by connecting electronic components in various ways. Exactly how you connect components together dictates how current flows through your circuit — and how voltage is dropped throughout the circuit.

Series connections

In the simple light-bulb circuit examined earlier in this chapter (refer to Figure 2-3), current flows from the positive battery terminal through the closed switch, through the light bulb, and then back to the negative terminal of the battery. You can call this arrangement a *series circuit,* which just means that the current runs through each component sequentially — in series.

Two important things you need to remember about series circuits are

- Each component has the same current.
- The voltage supplied by the source is divided (though not necessarily evenly) among the components. If you add up the voltage drops across each component, you get the total supply voltage.

There is a potential problem you may run into with series circuits: If one component fails, it creates an open circuit, stopping the flow of current to every component in the circuit. So, if your expensive new restaurant sign sports 200 light bulbs wired together in series to say "BEST FOOD IN TOWN," and a home-run ball knocks out one bulb, every one of the light bulbs goes dark.

Parallel connections

There's a way to fix the problem of all components in a series circuit blacking out when one component fails. You can wire the components using parallel connections — such as those in the circuit shown in Figure 2-5. With a parallel circuit, even if several baseballs take out a few bulbs in your sign, the rest of it stays lit. (Of course, you may be left with a glowing sign reading, "BEST FOO I OWN." There are pros and cons to everything.)

Here's how the parallel circuit in Figure 2-5 works: Current flows from the positive battery terminal, and then splits at each branch of the circuit, so each light bulb gets a share of the supply current. The current flowing through one light bulb doesn't flow through the other light bulbs. So, if your restaurant sign has 200 light bulbs wired together in parallel and one burns out, light still shines from the other 199 bulbs.

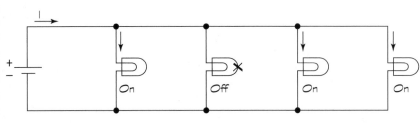

Figure 2-5: Lights bulbs are often arranged in a parallel circuit so if one burns out, the rest stay lit.

In parallel circuits, the voltage across each parallel branch of the circuit is the same. And when you know how to calculate the current flowing in each branch of the circuit (discussed in Chapter 3), you will see that if you add up all the branch currents, you get the total current supplied by the battery.

Two important things you need to remember about parallel circuits are

- ✔ Voltage across each branch is the same.
- ✔ Current supplied by the source is divided among the branches. The currents in each branch add up to the total supply current.

For the same circuit components, connecting them in parallel draws more current from your source than connecting them in series. If your circuit is powered by a battery, you need to be aware of just how long your battery can supply the necessary current to your circuit. As we discuss in Chapter 8, batteries have ratings of *amp-hours*. A battery with a rating of one amp-hour (for example) will last for just one hour in a circuit that draws one amp of current (theoretically, anyway; in practice, even new batteries don't always deliver on their amp-hour promises). Therefore, when you're deciding what power source to use for a circuit, you must take into account both the current that a circuit draws and how long you want to run the circuit.

Combination circuits

Most circuits are combinations of series and parallel connections. How you arrange components in a circuit depends on what you're trying to do. Look at the series-parallel circuit in Figure 2-6. You see a resistor (symbolized by a zigzag pattern in the diagram) in series with the battery, and then three parallel branches, each containing a switch in series with a light bulb. If all three switches are closed, the supply current travels through the resistor and then splits three different ways — with some current passing through each of the three bulbs. If all three switches are open, there is no complete path for current to follow, so no current flows out of the battery at all. If only one switch is closed, then all the supply current flows through just one bulb, and the other bulbs are off. By alternating which switch is open at any time, you can control which bulb is lit. You can imagine such a circuit controlling the operation of a three-stage traffic light (with a few more parts to control the timing and sequencing of the switching action).

To analyze combination circuits, you have to apply voltage and current "rules" one step at a time, using series rules for components in series, and parallel rules for components in parallel. At this point, you don't quite have enough information to calculate all the currents and voltages in the light bulb circuits we've shown here. You need to know about one more rule, called *Ohm's Law*, and then you'll have everything you need to analyze simple circuits. (We cover Ohm's Law and basic circuit analysis in Chapter 3.)

Figure 2-6:
By opening and closing switches in this series-parallel circuit, you can direct the supply current through different paths.

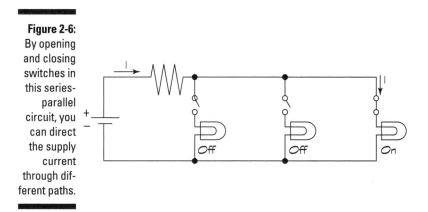

Creating Electronic Systems

To get an idea of how much you can accomplish by connecting various electronic components together in combination circuits, we take you on a tour of a couple of advanced electronic devices. But don't be worried; We don't expect you to follow the electrons as they traverse a complex web of circuitry. We just want to convince you that even ridiculously complicated electronic systems that consist of a mind-boggling assortment of components do the same sort of things that simple circuits do: manipulate electric current to perform a task.

Two advanced electronic systems are the radio-receiver system in your car and your television set.

Deciphering electrical signals

An *electrical signal* is the pattern over time of an electrical current. Often, the way an electrical signal changes its shape conveys information about something physical, such as the intensity of light, heat, or sound, or the position of an object, such as the diaphragm in a microphone or the shaft of a motor. Think of an electrical signal as a code, somewhat like Morse code, sending and receiving secret messages that you can figure out — if you know the key.

An *analog electrical signal,* or simply *analog signal,* is so named because it is an "analog," or one-to-one mapping, of the real physical quantity it represents. For instance, when a sound studio records a song, fluctuations in air pressure (that's what sound is) move the diaphragm

of a microphone, which produces corresponding variations in electrical current. That fluctuating current is a representation of the original sound, or an analog electrical signal.

Digital systems, like computers, can't handle continuous analog signals, so electrical signals must be converted into digital format before entering the depths of a digital system. *Digital format* is just another coding scheme which uses only the binary values 1 and 0 to represent information. (Hey! That's like Morse code's dot and dash!) A *digital signal* is created by sampling the value of an analog signal at regular intervals in time and converting each value into a string of *bits,* or binary digits.

Making sound appear out of thin air

Electronic components in a radio-receiver system control the current going to your speakers so you hear the sounds your favorite radio station transmits. As with most complex electronic systems, a radio-receiver system is made up of several stages, with each stage performing a specific function and the output of one stage feeding into the input of another stage. To get your speakers to reproduce the sounds originally created in the studio, the electronics in your stereo system performs these distinct functions:

- **Antenna:** Captures radio waves (invisible signals transmitted from many different radio stations) in the air and transforms them into an *electrical signal,* which is a variable flow of electric current. (See the sidebar, "Deciphering electrical signals.")

- **Tuner:** Picks out just one radio signal from all the signals captured by the antenna, rejecting all the others.

- **Detector:** Separates the audio signal (a replica of the original sound) from the radio signal (which, in addition to the audio signal, includes a "carrier" signal that transports the audio signal through the air).

- **Amplifier:** Boosts the tiny audio signal so you can hear it.

- **Speakers:** Transform the amplified audio signal into sound.

Each stage contains a bunch of electronic components combined in a way that manipulates or "shapes" the signal. After all that signal manipulation, the signal is sent to the speakers to be turned into sound. The specific pattern and intensity with which the electrical signal moves the diaphragm determine what exactly you're hearing and how loud it is.

Technical folks often use *block diagrams* to describe the functionality of complex electronic systems, like the radio receiver shown in Figure 2-7. Each block represents a circuit that takes the output of the preceding block as its input signal, performs some function, and produces an *output signal*, which is fed into another stage of the system.

Figure 2-7: Block diagram representing a radio-receiver system. The electronics in the system shape the electric current in several ways before it powers the speakers.

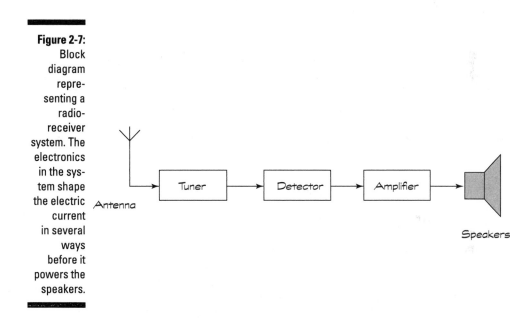

Painting pictures with electrons

Your television set, regardless of whether it uses an old-fashioned cathode-ray tube (CRT), newer plasma screen, or liquid-crystal display (LCD) to show you all those pictures, uses electronics to control which picture elements *(pixels)* get fired up on-screen at any given time to "paint" a picture with electrons. The electrical signal that enters your set carries information about the picture to be displayed (whether from your TV signal provider or the output of another electronic device, such as a DVD player). Electronic components within the set busily "decode" that electrical signal and apply the information carried by the signal to control the color and intensity of each pixel.

Different types of TVs activate display pixels in different ways. For example, the electronics in a color CRT TV "steer" three separate electron beams, positioning them to strike colored phosphors on the inside surface of the screen. The electronics also control whether each beam is "on" or "off" as it sweeps past each pixel. Result: The specific phosphor that the beam is aimed at either gets bombarded by electrons or is left alone. When a beam of electrons strikes a phosphor, it glows. By coordinating the movement and on/off state of the so-called "red," "blue," and "green" electron beams, the electronics in your TV create colorful images on the screen.

Chapter 3

Meeting Up with Resistance

· ·

In This Chapter

▶ Using resistance to your advantage

▶ Creating just the right amount of resistance with fixed and variable resistors

▶ Understanding how current, voltage, and resistance are governed by Ohm's Law

▶ Practicing Ohm's Law by analyzing circuits

▶ Using power as your guide in choosing circuit components

· ·

*I*f you toss a marble into a sandbox, the marble won't go very far. But if you toss a marble onto the surface of a large frozen lake, the marble will enjoy a nice little ride before it eventually comes to a stop. A mechanical force called friction stops that marble on either surface — it's just that the sand provides more friction than the ice.

Resistance in electronics is a lot like friction in mechanical systems: It puts the brakes on electrons (those teeny-tiny moving particles that make up electric current) as they move through materials.

This chapter looks at exactly what resistance is, where you can find resistance (everywhere), and how you can use it to your advantage by selecting *resistors* (components that provide controlled amounts of resistance) for your electronic circuits. Next you get a peek at the intimate relationship between voltage (the electrical force that pushes electrons) and current in components that have resistance, summed up quite nicely in a very simple equation with an authoritative name: Ohm's Law. The next order of business is to put Ohm's Law to work analyzing the goings-on in some basic circuits. Finally, you get a look at the role of Ohm's Law and related power calculations in the design of electronic circuits.

Resisting the Flow of Current

Resistance is a measure of an object's opposition to the flow of electrons. This may sound like a bad thing, but it's actually very useful. Resistance is what makes it possible to generate heat and light, restrict the flow of electric current when necessary, and ensure that the correct voltage is supplied to a device. For instance, as electrons travel through the filament of a light bulb, they meet so much resistance that they slow down a lot. As they fight their way through the filament, the atoms of the filament bump into each other furiously, generating heat — which produces the glow that you see from your light bulb.

Everything — even the best conductors — exhibits a certain amount of resistance to the flow of electrons. (Well, actually, there are certain materials, called *superconductors,* that can conduct current with zero electrical resistance — but only if you cool them down to extremely low temperatures. You won't encounter them in conventional electronics.) The higher the resistance, the more restricted the flow of current. So what determines how much resistance an object has? Resistance depends on several factors:

- **Material:** Some materials allow their electrons to roam freely, whereas others hold on tight to their electrons. Conductors have relatively low *resistivity,* whereas insulators have relatively high resistivity.

- **Cross-sectional area:** Resistance varies inversely with cross-sectional area; the larger the diameter, the easier it is for electrons to move — that is, the lower the resistance to their movement. Think of water flowing through a pipe: The wider the pipe, the easier the water flows. Along the same lines, a copper wire with a large diameter has a lower resistance than a copper wire with a small diameter.

- **Length:** The longer the material, the more resistance it has because there are more opportunities for electrons to bump into other particles along the way. Resistance varies directly with length.

- **Temperature:** For most materials, the higher the temperature, the higher the resistance. Higher temperatures mean that the particles inside have more energy, so they bump into each other a lot more, slowing down the flow of electrons. One notable exception to this is a type of resistor called a *thermistor:* Increase the temperature of a thermistor, and it *lowers* its resistance in a very predictable way. (You can imagine how useful that is in temperature-sensing circuits.) You can read about thermistors in Chapter 8.

You use the symbol, R, to represent resistance in an electronic circuit. Sometimes you'll see a subscript next to a resistance, for instance, R_{bulb}. That just means that R_{bulb} represents the resistance of the light bulb (or whatever part of the circuit the subscript refers to). Resistance is measured in units called *ohms,* abbreviated with the Greek letter, omega (Ω). The higher the ohm value, the higher the resistance.

A single ohm is so small a unit of resistance, that you're likely to see resistance measured in larger quantities, such as *kilohms* (kilo + ohm), which is thousands of ohms and is abbreviated $k\Omega$, or *megohms* (mega + ohm), which is millions of ohms and is abbreviated $M\Omega$. So 1 $k\Omega$ = 1,000 Ω and 1 $M\Omega$ = 1,000,000 Ω.

Resistors: Passive Yet Powerful

Resistors are passive electronic components that are specially designed to provide controlled amounts of resistance (for instance, 470 Ω or 1 $k\Omega$). Although a resistor won't provide gain or control the direction of current flow (because it's passive), you will find it to be a powerful little device because it enables you to put the brakes on current flow in a very controlled way. By carefully choosing and arranging resistors in different parts of your circuit, you can control just how much — or how little — current each part of your circuit gets.

What are resistors used for?

Resistors are among the most popular electronic components in town because they're simple yet versatile. One of the most common uses of a resistor is to limit the amount of current in part of a circuit, but resistors can also be used to control the amount of voltage provided to part of a circuit.

Limiting current

The circuit in Figure 3-1 shows a 9 V battery supplying current to a little device called a light-emitting diode (LED) through a resistor (shown as a zigzag). LEDs (like many other electronic parts) eat up current like a kid eats candy: They try to gobble up as much as you give them. But LEDs run into a problem — they burn themselves out if they draw too much current. The resistor in the circuit serves the very useful function of limiting the amount of current sent to the LED (the way a good parent restricts the intake of candy).

Too much current can destroy many sensitive electronic components — such as transistors (which we discuss in Chapter 6) and integrated circuits (which we discuss in Chapter 7). By putting a resistor at the input to a sensitive part, you limit the current that reaches the part. (But if you use too large a resistor, say 1 MK, you'll limit the current so much you won't see the light, although it's there!) This simple technique can save you a lot of time and money that you would otherwise lose fixing accidental blow-ups of your circuits.

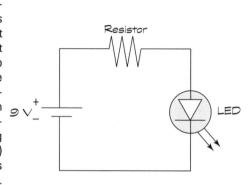

Figure 3-1:
The resistor limits the amount of current flowing into sensitive components, such as the light-emitting diode (LED) in this circuit.

Reducing and controlling voltage

Resistors can be used to reduce the voltage supplied to different parts of a circuit. Say, for instance, you have a 9 V power supply but you need to provide 5 V to power a particular integrated circuit you're using. You can set up a circuit, such as the one shown in Figure 3-2, to divide the voltage in a way that provides 5 V at the output. Then — *voilà* — you can use the output voltage, V_{out}, of this *voltage divider* as the supply voltage for your integrated circuit. (You can find details of exactly how this works later in this chapter.)

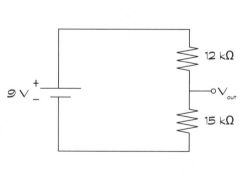

Figure 3-2:
Use two resistors to create a voltage divider, a common technique for producing different voltages for different parts of a circuit.

You can also put a resistor to work with another popular component — a capacitor, which we discuss in Chapter 4 — to create up and down voltage swings in a predictable way. You'll find the resistor-capacitor combo helps you create a kind of hourglass timer, which comes in handy for circuits that have time dependencies (for instance, a three-way traffic light). We show how the dynamic duo of resistor-and-capacitor operates in Chapter 4.

Choosing a type of resistor: Fixed or variable

Resistors come in two basic flavors: fixed and variable. Here's the lowdown on each type and why you would choose one or another:

- A *fixed resistor* supplies a constant, factory-determined resistance (no surprise there — that's what "fixed" means). But the actual resistance of any given resistor may vary (up or down) from its nominal value by some percentage, known as the *resistor tolerance.* So when you choose (say) a 1,000 Ω resistor with a 5 percent tolerance, the *actual* resistance it provides could be anywhere from 950 Ω to 1,050 Ω (because 5 percent of 1,000 is 50). Think of this as a resistance of 1,000 Ω, give or take 5 percent. There are two categories of fixed resistors (see Figure 3-3):

  - *High-precision resistors* come within just 1 percent of their nominal value. You use these in circuits where you need extreme accuracy, as in a precision timing or voltage reference circuit.

  - *Standard-precision resistors* can vary anywhere from 2 percent to (gulp) 20 percent of their nominal values. Markings on the resistor package will tell you just how far off the actual resistance may be (for instance, ±2%, ±5%, ±10%, or ±20%). You use standard-precision resistors in most hobby projects because (more often than not) you're using resistors to limit current or divide voltages to within an acceptable range. Resistors with 5 percent or 10 percent tolerance are commonly used in electronic circuits.

- A *variable resistor,* often called a *potentiometer* (*pot* for short) or *rheostat,* allows you to continually adjust the resistance from virtually zero ohms to a factory-determined maximum value. You use a potentiometer when you want to vary the amount of current or voltage you're supplying to part of your circuit. A few examples of where you'll find potentiometers: light-dimmer switches, volume controls for audio systems, and joystick controls (for games and on aircraft).

In circuit diagrams, also called *schematics* (detailed in Chapter 10), you use a zigzag symbol to represent a fixed resistor. There is no polarity indicator (+ or –) on a resistor; current is happy to flow either way through it. You add an arrow through the zigzag to create the schematic symbol for a *rheostat* (that is, a two-terminal variable resistor), and add an arrow pointing into the zigzag to create the symbol for a *potentiometer* (a three-terminal variable resistor). (See Figure 3-4.) We explain the difference between rheostats and potentiometers in the sidebar titled "What's in a name?" later in this chapter.

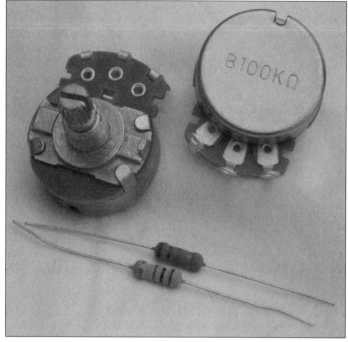

Figure 3-3:
Color-coded
bands
denote
the value
of a fixed
resistor;
potentiom-
eters are
usually
stamped
with their
highest
resistance
value.

Figure 3-4:
Resistor
symbols.

Fixed
resistor

Rheostat

Potentiometer

Reading into fixed resistors

Most fixed resistors come in a cylindrical package with two leads sticking out so you can connect them to other circuit elements (see the "Recognizing resistors on circuit boards" sidebar for exceptions). You'll be happy to know that you can insert a fixed resistor either way in a circuit — there's no left or right, up or down, to or from, when it comes to these pleasant little two-terminal devices.

The attractive rainbow colors adorning most resistors serve a purpose beyond catching your eye. Color coding identifies the *nominal value* and *tolerance* of most resistors, whereas others are drab and boring, and have their values stamped on them. The color code starts near the edge of one side of the resistor and consists of several stripes, or *bands,* of color. Each color represents a number, and the position of the band indicates how you use that number. Standard-precision resistors use four color bands: The first three bands indicate the nominal value of the resistor, and the fourth indicates the tolerance. High-precision resistors use five color bands: The first four bands indicate the value, and the fifth indicates the tolerance (typically ±1%).

Using a special decoder ring (just kidding — actually you read the resistor's color code, shown in the first two columns of Table 3-1), you can decipher the nominal value of a standard-precision resistor as follows:

✔ The **first band** gives you the first digit.

✔ The **second band** gives you the second digit.

✔ The **third band** gives you the multiplier as a number of zeros, except if it's gold or silver.

   • If the third band is **gold,** you multiply by 0.1 (divide by 10).

   • If the third band is **silver,** you multiply by 0.01 (divide by 100).

   Result: You get the nominal value of the resistance by putting the first two digits together (side by side) and applying the multiplier.

The fourth band (tolerance) uses a different color code, as shown in the third column of Table 3-1. If there is no fourth band, you can assume the tolerance is ±20%.

Table 3-1	Resistor Color Coding	
Color	*Number*	*Tolerance*
Black	0	±20%
Brown	1	±1%
Red	2	±2%
Orange	3	±3%
Yellow	4	±4%
Green	5	n/a
Blue	6	n/a
Violet	7	n/a
Gray	8	n/a
White	9	n/a
Gold	0.1	±5%
Silver	0.01	±10%

Take a look at a couple of examples:

- ✔ **Red-red-yellow-gold:** A resistor with red (2), red (2), yellow (4 zeros), and gold (±5%) bands represents a nominal resistance of 220,000 Ω, or 220kΩ, which could vary up or down by as much as 5 percent of that value. So it could have a resistance of anywhere between 209kΩ and 231kΩ.

- ✔ **Orange-white-gold-silver:** A resistor with orange (3), white (9), gold (0.1), and silver (±10%) bands represents a value of 39 × 0.1, or 3.9 Ω, which could vary by up to 10 percent of that value. So the actual resistance could be anywhere from 3.5 Ω to 4.3 Ω.

For high-precision resistors, the first three bands of color give you the first three digits, the fourth band gives you the multiplier, and the fifth band represents the tolerance.

Most circuit designs tell you the safe resistor tolerance to use, whether for each individual resistor or for all the resistors in the circuit. Look for a notation in the parts list or as a footnote at the bottom of the circuit diagram. If the schematic doesn't state a tolerance, then you can assume it's okay to use standard tolerance resistors (±5% or ±10%).

Dialing with potentiometers

Potentiometers consist of a resistance track with connections at both ends and a *wiper* that moves along the track as you adjust the resistance across a range from 0 (zero) Ω to some maximum value (see Figure 3-5). Most often, potentiometers are marked with their maximum value — 10k, 50k, 100k, 1M, and so forth — and they don't always include the little ohm symbol (Ω). For example, with a 50k pot, you can dial in any resistance from 0 to 50,000 Ω.

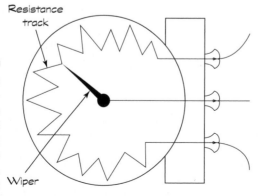

Figure 3-5:
A potentiometer has a wiper that moves along a resistance track.

What's in a name?

The word *potentiometer* is often used to categorize all variable resistors, but there is a difference between rheostats and potentiometers. Rheostats are two-terminal devices, with one lead connected to the wiper and the other lead connected to one end of the resistance track. Technically, a potentiometer is a three-terminal device; its leads connect to the wiper and to *both* ends of the resistance track. You can use a potentiometer as a rheostat (as is quite common) by connecting only two of its leads, or you can connect all three leads in your circuit — and get both a fixed and variable resistor for the price of one!

Rheostats typically handle higher levels of voltage and current than potentiometers. This makes them ideal for industrial applications, such as controlling the speed of electric motors in large machines. However, rheostats have largely been replaced by circuits that use semiconductor devices (see Chapter 6), which consume much less power.

Bear in mind that the range on the potentiometer is approximate only. If the potentiometer lacks markings, use a multimeter to figure out the component's value. (Chapter 12 shows how to test resistances using a multimeter.)

Potentiometers allow you to adjust resistance continuously, and are available in various packages known as dial pots, slide pots, and trim pots:

- *Dial pots* contain rotary resistance tracks and are controlled by turning a shaft or knob. Commonly used in electronics projects, dial pots are designed to be mounted through a hole cut in a case that houses a circuit, with the knob accessible from the outside of the case. Dial pots are popular for adjusting volume in sound circuits.

- *Slide pots* contain a linear resistance track and are controlled by moving a slide along the track. You see them on stereo equipment and some dimmer switches.

- *Trim pots* (also known as *preset pots*) are smaller, are designed to be mounted on a circuit board, and provide a screw for adjusting resistance. They are typically used to fine-tune a circuit design — for instance, to set the sensitivity of a light-sensitive circuit — rather than to allow for variations (such as volume adjustments) during the operation of a circuit.

If you use a potentiometer in a circuit, bear in mind that if the wiper is "dialed down" all the way, you've got zero resistance, and you aren't limiting current at all with this device. It's common practice to insert a fixed resistor in series with a potentiometer as a "safety net" to limit current. You just choose a value for the fixed resistor so it works together with your variable resistor to produce the range of resistance you need. (Look for details later in this chapter about figuring out the total resistance of multiple resistors in series.)

Recognizing resistors on printed circuit boards

As you learn more about electronics, you may get curious enough to take a look inside some of the electronics in your house. (Warning: Be careful! Follow the safety guidelines given in Chapter 9.) You might (for example) open up the remote control for your TV and see some components wired up between a touchpad and an LED. On some *printed circuit boards (PCBs)* — which serve as platforms for building the mass-produced circuits commonly found in computers and other electronic systems — you may have trouble recognizing the individual circuit components. That's because manufacturers use fancy

techniques to populate PCBs with components, aiming to eke out efficiencies and save space on the boards. One such technique, *surface-mount technology (SMT),* allows components to be mounted directly to the surface of a board (think of them as "hitting the deck"). *Surface-mount devices,* such as SMT resistors, look a bit different from the components you would use to build a circuit in your garage, because they don't require long leads in order to connect them within a circuit. Such components use their own coding system to label the value of the part.

You can always use a multimeter to measure the resistance of an unknown resistor or the variable resistance provided by a potentiometer. (See Chapter 12 for the how-to details.)

Rating resistors according to power

Quiz time! What do you get when you let too many electrons pass through a resistor at the same time? If you answered "a charred mess and no money-back guarantee," you're right! Whenever electrons flow through something with resistance, they generate heat — and the more electrons, the higher the heat. Electronic components (such as resistors) can only stand so much heat (just how much depends on the size and type of component) before they have a meltdown. Because heat is a form of energy, and power is a measure of the energy consumed over a period of time, you can use the *power rating* of an electronic component to tell you how many watts (what? *Watts,* abbreviated W, are units of electric power) a component can safely handle.

All resistors come with power ratings. Standard, run-of-the-mill resistors can handle $1/8$ W or $1/4$ W, but you can easily find $1/2$ W and 1 W resistors — and some are even flameproof. (Does that make you nervous about building circuits?) Of course, you won't see the power rating indicated on the resistor itself (that would make it too easy), so you have to figure it out by the size of the resistor (the bigger the resistor, the more power it can handle) or get it from the manufacturer or your parts supplier.

So how do you use the power rating to choose a particular resistor for your circuit? You estimate the peak power that your resistor will be expected to handle, and pick a power rating that meets or exceeds it. Power is calculated as follows:

$$P = V \times I$$

V represents the voltage (in volts, abbreviated V) measured across the resistor and I represents the current (in amps, abbreviated A) flowing through the resistor. For example, suppose the voltage is 5 V and you want to pass 25 mA (milliamps) of current through the resistor. To calculate the power, you multiply 5 by 0.025 (remember, *milli*amps are thousandths of amps) and you get 0.125 W, or $^{1}/_{8}$ W. So you know that a $^{1}/_{8}$ W resistor may be okay, but you can be sure that a $^{1}/_{4}$ W resistor *will* take the heat just fine in your circuit.

For most hobby electronics projects, $^{1}/_{4}$ W or $^{1}/_{8}$ W resistors are A-OK. You need high-wattage resistors for *high-load* applications, where loads, such as a motor or lamp control, require higher-than-hobby-level currents to operate. High-wattage resistors take many forms, but you can bet they are bigger and bulkier than your average resistor. Resistors with power ratings of over 5 W are gift-wrapped in epoxy (or other waterproof and flameproof coating) and have a rectangular, rather than cylindrical, shape. A high-wattage resistor may even include its own metal *heat sink,* with fins that conduct heat away from the resistor.

Combining Resistors

When you start shopping for resistors, you'll find that you can't always get exactly what you want. That's because it would be impractical for manufacturers to make resistors with every possible value of resistance. So they make resistors with a limited set of resistance values, and you work around it (as you're about to see). For instance, you can search far and wide for a 25kΩ resistor, but you may never find it; however, 22kΩ resistors are as common as the day is long! The trick is to figure out how to get the resistance you need using standard available parts.

As it turns out, you can combine resistors in various ways to create an *equivalent resistance* value that will come pretty darn close to whatever resistance you need. And because standard precision resistors are accurate to 5 or 10 percent of their nominal value anyway, combining resistors works out just fine.

There are certain "rules" for combining resistances, which we cover in this section. Use these rules not only to help you choose off-the-shelf resistors for your own circuits, but also as a key part of your effort to analyze other people's electronic circuits. For instance, if you know that a light bulb has a certain amount of resistance, and you place a resistor in series with the bulb to limit the current, you'll need to know what the total resistance of the two components is before you can calculate the current passing through them.

Resistors in series

When you combine two or more resistors (or resistances) in series, you connect them end to end (as shown in Figure 3-6), so that the same current passes sequentially through each resistor. By doing this, you restrict the current somewhat with the first resistor, you restrict it even more with the next resistor, and so forth. So the effect of the series combination is an *increase* in the overall resistance.

To calculate the combined (equivalent) resistance of multiple resistors in series, you simply add up the values of the individual resistances. You can extend this rule to any number of resistances in series:

$$R_{series} = R1 + R2 + R3 + R4 + \ldots$$

R1, R2, R3, and so forth represent the values of the resistors and R_{series} represents the total equivalent resistance. Remember that the same current flows through all resistors connected in series.

You can apply this concept of equivalent resistance to help you select resistors for a specific circuit need. For example, suppose you need a 25kΩ resistor, but cannot find a standard resistor with that value. You can combine two standards resistors — a 22kΩ resistor and a 3.3kΩ resistor — in series to get 25.3kΩ of resistance. That's less than 2 percent different from the 25kΩ you seek — well within typical resistor tolerance levels (which are 5–10 percent).

Be careful with your units of measurement when you add up resistance values. For example, suppose you connect the following resistors in series (see Figure 3-6): 1.2kΩ, 680 Ω, and 470 Ω. Before you add the resistances, you need to convert the values to the same units — for instance, ohms. In this case, the total resistance, R_{total}, is calculated as follows:

$$R_{total} = 1,200 \ \Omega + 680 \ \Omega + 470 \ \Omega = 2,350 \ \Omega \text{ or } 2.35\text{k}\Omega$$

Figure 3-6:
The combined resistance of two or more resistors in series is the sum of the individual resistances.

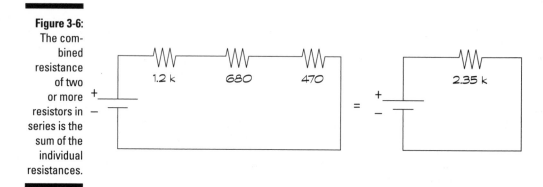

The combined resistance will *always* be greater than any of the individual resistances. This fact comes in handy when you're designing circuits. For example, if you want to limit current going into a light bulb, but you don't know the resistance of the bulb, you can place a resistor in series with the bulb and be secure in the knowledge that the total resistance to current flow is *at least* as much as the value of the resistor you added. For circuits that use variable resistors (such as a light-dimmer circuit), putting a fixed resistor in series with the variable resistor guarantees that the current will be limited even if the pot is dialed down to zero Ω. (The lowdown on how to calculate just what the current will be for a given voltage/resistance combo appears later in this chapter.)

Resistors in parallel

When you combine two resistors in parallel, you connect both sets of ends together (see Figure 3-7) so each resistor has the same voltage. By doing so, you provide two different paths for current to flow, so even though each resistor is restricting current flow through one circuit path, there's still another path that can draw additional current. From the perspective of the source voltage, the effect of arranging resistors in parallel is a *decrease* in the overall resistance.

To calculate the equivalent resistance, $R_{parallel}$, of two resistors in parallel, you use the following formula:

$$R_{parallel} = \frac{R1 \times R2}{R1 + R2}$$

where R1 and R2 are the values of the individual resistors.

In the example in Figure 3-7, two 2kΩ resistors are placed in parallel. The equivalent resistance is as follows:

$$R_{parallel} = \frac{2,000 \times 2,000}{2,000 + 2,000}$$
$$= \frac{4,000,000}{4,000}$$
$$= 1,000$$

$$R_{parallel} = 1k\Omega$$

In this example, because the two resistors have equal resistance, connecting them in parallel results in an equivalent resistance of *half the value of either one*. The result is that each resistor draws half the supply current. If two resistors of unequal value are placed in parallel, *more* current will flow through the path with the *lower* resistance than the path with the higher resistance.

Figure 3-7:
The combined resistance of two or more resistors in parallel is always lower than any of the individual resistances.

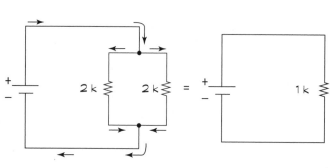

If your circuit calls for a resistor with a somewhat higher power rating, say 1 W, but you only have $^1/_2$ W resistors on hand, you can combine two $^1/_2$ W resistors in parallel instead. Just select resistor values that combine to create the resistance you need. Since each one draws half the current that a single resistor would draw, it dissipates half the power (remember that power = current × voltage).

If you combine more than one resistor in parallel, the math gets a little more complicated:

$$R_{parallel} = \cfrac{1}{\cfrac{1}{R1} + \cfrac{1}{R2} + \cfrac{1}{R3}}\cdots(\text{and more as needed})$$

For multiple resistances in parallel, the amount of current flowing through any given branch is *inversely proportional* to the resistance within that branch. In practical terms, the higher the resistance, the less current goes that way; the lower the resistance, the more current goes that way. Just like water, electric current favors the path of least resistance.

As a shorthand in electronic equations, you may see the symbol $||$ used to represent the formula for resistors in parallel. For example:

$$R_{parallel} = R1\|R2 = \frac{R1 \times R2}{R1 + R2}$$

or

$$R_{parallel} = R1\|R2\|R3 = \cfrac{1}{\cfrac{1}{R1} + \cfrac{1}{R2} + \cfrac{1}{R3}}$$

Combining series and parallel resistors

Many circuits combine series resistors and parallel resistors in various ways so as to restrict current in some parts of the circuit while splitting current in other parts of the circuit. In some cases, you can calculate equivalent resistance by combining the equations for resistors in series and resistors in parallel. For instance, in Figure 3-8, resistor R2 (2kΩ) is in parallel with resistor R3 (2kΩ), and that parallel combination is in series with resistor R1 (1kΩ). You can calculate the total resistance (in kΩ) as follows:

$$R_{total} = R1 + \left(R2 \| R3 \right)$$
$$= R1 + \frac{R2 \times R3}{R2 + R3}$$
$$= 1 + \frac{2 \times 2}{2 + 2}$$
$$= 1 + 1$$
$$= 2$$
$$R_{total} = 2k\Omega$$

In this circuit, the current supplied by the battery is limited by the *total* resistance of the circuit, which is 2kΩ. Supply current flows from the positive battery terminal through resistor R1, splits — with half flowing through resistor R2 and half flowing through resistor R3 — and then combines again to flow into the negative battery terminal.

Figure 3-8:
Many circuits include a combination of parallel and series resistances.

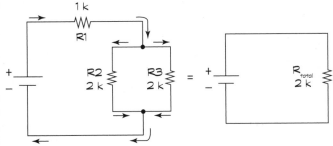

Circuits often have more complex combinations of resistances than simple series or parallel relationships, and figuring out equivalent resistances isn't always easy. You have to use matrix mathematics to analyze them, and because this book isn't a math book, we're not going to detour to explore the complexities of matrix math.

Obeying Ohm's Law

One of the most important concepts to understand in electronics is the relationship between voltage, current, and resistance in a circuit, summed up in a simple equation known as Ohm's Law. When you understand this thoroughly, you will be well on your way to analyzing circuits that other people have designed, as well as successfully designing your own circuits. Before diving into Ohm's Law, it may help to take a quick look at the ebbs and flows of current.

Driving current through a resistance

If you place a voltage source across an electronic component that has measurable resistance (such as a light bulb or a resistor), the force of the voltage will push electrons through the component. The movement of gobs of electrons is what constitutes electric current. By applying a greater voltage, you exert a stronger force on the electrons, which creates a stronger flow of electrons — a larger current — through the resistance. The stronger the force (voltage V), the stronger the flow of electrons (current I).

This is analogous to water flowing through a pipe of a certain diameter. If you exert a certain water pressure on the water in the pipe, the current will flow at a certain rate. If you increase the water pressure, the current will flow faster through that same pipe, and if you decrease the water pressure, the current will flow slower through the pipe.

It's constantly proportional!

The relationship between voltage (V) and current (I) in a component with resistance (R) was discovered in the early 1800s by Georg Ohm (does his name sound familiar?). He figured out that for components with a fixed resistance, voltage and current vary in the same way: Double the voltage, and the current is doubled; halve the voltage, and the current is halved. He summed this relationship up quite nicely in the simple mathematical equation that bears his name: Ohm's Law.

Ohm's Law states that voltage equals current multiplied by resistance, or

$$V = I \times R$$

What this really means is that the voltage (V), measured across a component with a fixed resistance, is equal to the current (I) flowing through the component multiplied by the value of the resistance (R).

For example, in the simple circuit in Figure 3-9, a 9 V battery applied across a 1kΩ resistor produces a current of 9 mA (which is 0.009 A) through the circuit:

$$9 \text{ V} = 1{,}000 \text{ } \Omega \times 0.009 \text{ A}$$

This little law is so important in electronics that you'd be wise to repeat it over and over again, like a mantra, until you've been transformed into an Ohm's Law-abiding geek! To help you remember, think of Ohm's Law as a <u>V</u>ery <u>I</u>mportant <u>R</u>ule.

Figure 3-9:
A voltage of
9 V applied
to a resistor
of 1kΩ
produces a
current of
9 mA.

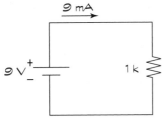

When using Ohm's Law, watch your units of measurement carefully. Make sure you convert any *kilos* and *millis* before you get out your calculator. If you think of Ohm's Law as *volts = amps × ohms,* you'll be okay. And if you're brave, you can also use *volts = milliamps × kilohms,* which works just as well (because the *millis* cancel out the *kilos*). But if you aren't careful and you mix units, you may be in for a shock! For instance, a lamp with a resistance of 100 Ω passes a current of 50 mA. If you forget to convert milliamps to amps, you'll multiply 100 by 50 to get 5,000 V as the voltage across the lamp! Ouch! The correct way to perform the calculation is to convert 50 mA to 0.05 A, and *then* multiply by 100 Ω, to get 5 V. Much better!

There's a reason Georg Ohm has his name associated with resistance values as well as "the law." The way an ohm, or unit of resistance, is defined came from Georg Ohm's work. The *ohm* is defined as the resistance between two points on a conductor when one volt, applied across those points, produces one amp of current through the conductor. We just thought you might like to know that. (Good thing Georg's last name wasn't Wojciehowicz!)

One law, three equations

Remember your high-school algebra? Remember how you can re-arrange the terms of an equation containing variables (such as the familiar *x* and *y*) to solve for one variable, as long as you know the values of the other variables? Well, the same rules apply to Ohm's Law. You can rearrange its terms to create two more equations, for a total of three equations from that one law!

$$V = I \times R \qquad I = {}^V/_R \qquad R = {}^V/_I$$

These three equations all say the same thing, but in different ways. You can use them to calculate one quantity when you know the other two. Which one you use at any given time depends on what you're trying to do. For example:

- **To calculate an unknown voltage,** multiply the current times the resistance ($V = I \times R$). For instance, if you have a 2 mA current running through a 2kΩ resistor, the voltage across the resistor is 2 mA × 2kΩ (or 0.002 A × 2,000 Ω) = 4 V.

- **To calculate an unknown current,** take the voltage and divide it by the resistance ($I = {}^V/_R$). For example, if 9 V is applied across a 1kK resistor, the current is 9 V/1000 Ω = 0.009 A or 9 mA.

- **To calculate an unknown resistance,** take the voltage and divide it by the current ($R = {}^V/_I$). For instance, if you have 3.5 V across an unknown resistor with 10 mA of current running through it, the resistance is ${}^{3.5V}/_{0.01A} = 350$ Ω.

Using Ohm's Law to Analyze Circuits

When you've got a good handle on Ohm's Law, you'll be ready to put it into practice. Ohm's Law is like a master key, unlocking the secrets to electronic circuits. Use it to understand circuit behavior and to track down problems within a circuit (for instance, why the light isn't shining, the buzzer isn't buzzing, or the resistor isn't resisting because it melted). You can also use it to design circuits and pick the right parts for use in your circuits. We get to that in the next section of this chapter. In this section, we discuss how to apply Ohm's Law to analyze circuits.

Calculating current through a component

In the simple circuit in Figure 3-10, a 6 V battery is applied across a 1kΩ resistor. You calculate the current through the resistor as follows:

$$I = {}^{6V}/_{1,000}\,\Omega = 0.006 \text{ A or 6 mA}$$

Figure 3-10:

Figure 3-10:
Calculating
the current
through the
resistor in
this simple
circuit is
a straight-
forward
application
of Ohm's
Law.

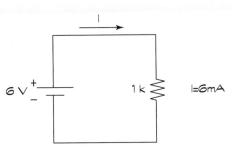

If you add a 220 Ω resistor in series with the 1kΩ, as shown in Figure 3-11, you're restricting the current even more. To calculate the current flowing through the circuit, you need to determine the total resistance that the 6 V battery is facing in the circuit. Because the resistors are in series, the resistances add up, for a total equivalent resistance of 1.22kΩ. The new current is:

$$I = {}^{6V}/_{1,220}\,\Omega \approx 0.0049\ \text{A or}\ 4.9\ \text{mA}$$

By adding the extra resistor, you've reduced the current in your circuit from 6 mA to 4.9 mA.

TIP

That double squiggle symbol (≈) in the equation just given means "is approximately equal to," and we used that because we rounded off the current to the nearest tenth of a milliamp. It's usually okay to round off the tinier parts of values in electronics — unless you're working on the electronics that control a sub-atomic particle smasher or other high-precision industrial device.

Figure 3-11:
Calculating
the current
through
this circuit
requires
determin-
ing the
equivalent
resistance,
and then
applying
Ohm's Law.

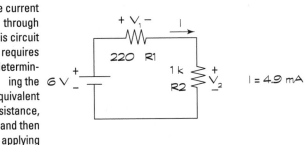

Calculating voltage across a component

In the circuit in Figure 3-10, the voltage across the resistor is simply the voltage supplied by the battery: 6 V. That's because the resistor is the only circuit element other than the battery. Adding a second resistor in series (as in Figure 3-11) changes the voltage picture: Now *some* of the battery voltage is dropped across the 1kΩ resistor (R2) and the *rest* of the battery voltage is dropped across the 220 Ω resistor (R1).

To figure out how much voltage is dropped across each resistor, you use Ohm's Law for each individual resistor. You know the value of each resistor, and *you know the current flowing through each resistor*. Remember, that current (I) is the battery voltage (6 V) divided by the total resistance (R1 + R2, or 1.22 kΩ), or approximately 4.9 mA. Now you can apply Ohm's Law to each resistor to calculate its voltage drop:

$$V_1 = I \times R1$$
$$= 0.0049 \text{ A} \times 220 \text{ }\Omega = 1.078 \text{ V} \approx 1.1 \text{ V}$$
$$V2 = I \times R2$$
$$= 0.0049 \text{ A} \times 1,000 \text{ }\Omega = 4.9 \text{ V}$$

Notice that if you add up the voltage drops across the two resistors, you get 6 volts, which is the total voltage supplied by the battery. That isn't a coincidence; the battery is supplying voltage to the two resistors in the circuit, and the supply voltage is divided between the resistors proportionally, according to the values of the resistors. This type of circuit is known as a *voltage divider*.

Many electronic systems use voltage dividers to bring down a supply voltage to a lower level, after which they feed that reduced voltage into the input of another part of the overall system that requires that lower voltage.

There's a quicker way to calculate either of the "divided voltages" (V_1 or V_2) in Figure 3-11. You know that the current passing through the circuit can be expressed as:

$$I = \frac{V_{battery}}{R1 + R2}$$

You also know that:

$$V_1 = I \times R1$$

and

$$V_2 = I \times R2$$

Using Ohm's Law

Ohm's Law is very useful in analyzing voltage and current for resistors and other components that behave like resistors, such as light bulbs. But you have to be careful about applying Ohm's Law to other electronic components, such as capacitors (detailed in Chapter 4) and inductors (covered in Chapter 5), which don't exhibit a *constant* resistance under all circumstances. For such components, the opposition to current — known as *impedance* — can and will vary depending on what's going on in the circuit. So you can't just use a multimeter to measure the "resistance" of a capacitor, for instance, and then try to apply Ohm's Law willy-nilly. For a closer look at impedance, check out Chapter 4.

To calculate V_1, for example, you can substitute the equation for I shown above, and you get:

$$V_1 = \frac{V_{battery}}{R1 + R2} \times R1$$

You can rearrange the terms, without changing the equation, to get:

$$V_1 = \frac{R1}{R1 + R2} \times V_{battery}$$

Similarly, the equation for V_2 is:

$$V_2 = \frac{R2}{R1 + R2} \times V_{battery}$$

By plugging in the values of R1, R2, and $V_{battery}$, you get $V_1 = 1.078$ V and $V_2 = 4.9$ V, just as calculated earlier.

You will see this general equation commonly used for the voltage across a resistor (R1) in a voltage divider circuit:

$$V_1 = \frac{R1}{R1 + R2} \times V_{battery}$$

Use the voltage divider equation to calculate the output voltage, V_{out}, in the voltage divider circuit shown in Figure 3-2, as follows:

$$V_{out} = \frac{15,000\ \Omega}{\left(12,000 + 15,000\right)\ \Omega} \times 9\text{ V}$$

$$= \frac{15,000}{27,000} \times 9\text{ V}$$

$$= 5\text{ V}$$

The circuit in Figure 3-2 divides a 9 V supply down to 5 V.

Calculating an unknown resistance

Say you have a large flashlight that you're running off a 12 V battery, and you measure a current of 1.3 A through the circuit (we discuss how to measure current in Chapter 12). You can calculate the resistance of the bulb by taking the voltage across the bulb (12 V) and dividing it by the current through the bulb (1.3 A). It's a pretty quick number-crunch:

$$R_{bulb} = {}^{12\,V}/_{1.3\,A} = 9\ \Omega$$

What Is Ohm's Law Really Good For?

Ohm's Law also comes in handy when you're analyzing all kinds of circuits, whether simple or complex. You'll use it in designing and altering electronic circuits, to make sure you get the right current and voltage to the right places in your circuit. You'll use Ohm's Law so much, it will become second nature to you.

Analyzing complex circuits

Ohm's Law really comes in handy when analyzing more complex circuits than the simple light bulb circuit we discussed earlier. You often need to incorporate your knowledge of equivalent resistances in order to apply Ohm's Law and figure out exactly where current is flowing and how voltages are being dropped throughout your circuit.

For example, let's say you add to the circuit in Figure 3-11 by placing a 2.2kΩ resistor in parallel with the 1kΩ resistor, as shown in Figure 3-12. You can calculate the current running through each resistor, step by step, as follows:

1. **Calculate the equivalent resistance of the circuit.**

   You can find this value by applying the rules for resistors in parallel and resistors in series, like this:

   $$R_{equivalent} = 220 + \frac{1,000 \times 2,200}{1,000 + 2,200}$$

   $$\approx 220 + 688$$

   $$\approx 908\ \Omega$$

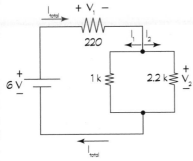

Figure 3-12:
Complex
circuits can
be analyzed
by applying
Ohm's
Law and
calculating
equivalent
resistances.

2. Calculate the total current supplied by the battery.

Here you apply Ohm's Law, using the battery voltage and the equivalent resistance:

$$I_{total} = {}^{6\,V}/_{908\,\Omega}$$

$$\approx 0.0066 \text{ A or } 6.6 \text{ mA}$$

3. Calculate the voltage dropped across the parallel resistors.

You can do this either of two ways:

- *Apply Ohm's Law to the parallel resistors.* You calculate the equivalent resistance of the two resistors in parallel, and then multiply that by the supply current. The equivalent resistance is 688 Ω, as shown in the first step above. So the voltage is

$$V_2 = 0.0066 \text{ A} \times 688 \text{ } \Omega$$

$$\approx 4.55 \text{ V}$$

- *Apply Ohm's Law to the 220 Ω resistor, and subtract its voltage from the supply voltage.* The voltage across the 220 Ω resistor is

$$V_1 = 0.0066 \text{ A} \times 220 \text{ } \Omega$$

$$\approx 1.45 \text{ V}$$

So the voltage across the parallel resistors is

$$V_2 = V_{supply} - V_1$$

$$= 6 \text{ V} - 1.45 \text{ V} = 4.55 \text{ V}$$

4. **Finally, calculate the current through each parallel resistor.**

   To get that result, you apply Ohm's Law to each resistor, using the voltage you just calculated (V_2). Here's what it looks like:

   $I_1 = {}^{4.55\,V}/_{1,000\,\Omega} \approx 0.0046$ A or 4.6 mA

   $I_2 = {}^{4.55\,V}/_{2,200\,\Omega} \approx 0.002$ A or 2 mA

   Notice that the two branch currents, I_1 and I_2, add up to the total supply current, I_{total}: 4.6 mA + 2 mA = 6.6 mA. That's a good thing (and a good way to check that you've done your calculations correctly).

Designing and altering circuits

You can use Ohm's Law to determine what components to use in a circuit design. For instance, you may have a series circuit consisting of a 9 V power supply, a resistor, and an LED, as shown in Figure 3-1 at the beginning of this chapter. As you see in Chapter 6, the voltage drop across an LED remains constant for a certain range of current passing through it, but if you try to pass too much current through the LED, it will burn out. Say (for example) your LED voltage is 2 V and the maximum current it can handle is 25 mA. What resistance should you put in series with the LED to limit the current so it never exceeds 25 mA?

To figure this out, first you have to calculate the voltage drop across the resistor when the LED is on. You already know that the supply voltage is 9 V and the LED eats up 2 V. The only other component in the circuit is the resistor, so you know that it will eat up the remaining supply voltage — all 7 V of it. If you want to limit the current to be no more than 25 mA, you need a resistor that is *at least* ${}^{7\,V}/_{0.025\,A} = 280$ Ω. Because you can't find a 280 Ω resistor, suppose you choose a 300 Ω resistor. The current will be ${}^{7\,V}/_{300\,\Omega} = 0.023$ A or 23 mA. The LED may burn a little less brightly, but that's okay.

Ohm's Law also comes in handy when tweaking an existing circuit. Say your spouse is trying to sleep but you want to read, so you get out your big flashlight. The bulb in your flashlight has a resistance of 9 Ω and is powered by a 6 V battery, so you know that the current in the flashlight circuit is ${}^{6\,V}/_{9\,\Omega} = 0.65$ A. Your spouse thinks the light is too bright, so to reduce the brightness (and save your marriage), you want to restrict the current flowing through the bulb a bit. You think that bringing it down to 0.45 A will do the trick, and you know that inserting a resistor in series between the battery and the bulb will restrict the current. But what value of resistance do you need? You can use Ohm's Law to figure out the resistance value as follows:

✔ Using the desired new current, calculate the desired voltage drop across the bulb: $V_{bulb} = 0.45$ A $\times$ 9 Ω ≈ 4.1 V

✔ Calculate the portion of the supply voltage you'd like to apply across the new resistor. This voltage is the supply voltage less the voltage across the bulb: $V_{resistor}$ = 6 V – 4.1 V = 1.9 V

✔ Calculate the resistor value needed to create that voltage drop given the desired new current. R = $^{1.9\,V}/_{0.45\,A}$ ≈ 4.2 Ω

✔ Finally, choose a resistor value that is close to the calculated value, and make sure it can handle the power dissipation: $P_{resistor}$ = 1.9 V × 0.45 A ≈ 0.9 W

Result: Because you won't find a 4.2 Ω resistor, you can use a 4.7 Ω 1 W resistor to reduce the brightness of the light. Your spouse can sleep so soundly; let's hope the snoring won't interfere with your reading!

The Power of Joule's Law

Another scientist hard at work in the early 1800s was the energetic James Prescott Joule. Joule is responsible for coming up with the equation that gives you power values (discussed earlier in this chapter); it's known as *Joule's Law:*

$$P = V \times I$$

This equation states that the power (in watts) equals the voltage (in volts) across a component times the current (in amps) passing through that component. The really nice thing about this equation is that it applies to every electronic component, whether it's a resistor, a light bulb, a capacitor, or something else. It tells you the rate at which electrical energy is consumed by the component — what that power is.

Using Joule's Law to choose components

You've already seen how to use Joule's Law to ensure a resistor is big enough to resist a meltdown in a circuit, but you should know that this equation also comes in handy when you're selecting other electronic parts. Lamps, diodes (as discussed in Chapter 6), and other components also come with maximum power ratings. If you expect them to perform at power levels higher than their ratings, you're going to be disappointed when too much power makes them pop and fizzle. When you select the part, you should consider the *maximum possible* power the part will need to handle in the circuit. You do this by determining the maximum current you'll be passing through the part and the voltage across the part, and then multiplying those quantities together. Then you choose a part with a power rating that exceeds that estimated maximum power.

Joule and Ohm: perfect together

You can get creative and combine Joule's Law and Ohm's Law to derive more useful equations to help you calculate power for resistive components and circuits. For instance, if you substitute $I \times R$ for V in Joule's Law, you get:

$$P = (I \times R) \times I = I^2 R$$

That gives you a way to calculate power if you know the current and resistance but not the voltage. Similarly, you can substitute V/R for I in Joule's Law to get

$$P = V \times {}^V\!/_R = {}^{V^2}\!/_R$$

Using that formula, you can calculate power if you know the voltage and the resistance but not the current.

Joule's Law and Ohm's Law are used in combination so often that Georg Ohm sometimes gets the credit for both laws!

Trying Your Hand at Circuits with Resistors

If you want to get your hands dirty and experiment with some real resistor circuits, you can take a look at some of the practice circuits in the beginning of Chapter 14 in Part III. These circuits show you Ohm's Law in action — and you can tweak a potentiometer to vary resistance and divide up voltages. But before you jump that far ahead, we recommend that you read through Part II, which explains how to set up your electronics shop, stay safe, read schematics, construct circuits, and measure everything in sight.

Chapter 4

Getting a Charge Out of Capacitors

● ●

In This Chapter

▶ Storing electrical energy in capacitors

▶ Charging and discharging capacitors

▶ Saying "no" to DC and "yes" to AC

▶ Seeing how capacitors react to different frequencies

▶ Using Ohm's Law (carefully!) for capacitive circuits

▶ Creating the dynamic duo: capacitors and resistors

▶ Exploiting capacitors to block, filter, smooth, and delay signals

● ●

*I*f resistors are the most popular electronic component, then capacitors are a close second. Skilled at storing electrical energy, capacitors are important contributors to all sorts of electronic circuits — and your life would be a lot duller without them.

Capacitors make it possible to change the shape (the pattern over time) of electrical signals carried by current — a task resistors alone cannot perform. Although they're not as straightforward to understand as resistors, capacitors are essential ingredients in many of the electronic and industrial systems you enjoy today, such as radio receivers, computer memory devices, and automobile airbag-deployment systems, so it's well worth investing your time and brain power in understanding how capacitors operate.

This chapter looks at what capacitors are made of, how they store electrical energy, and how circuits use that energy. You get a look at a capacitor charging up and later releasing its energy, and observe how it reacts to signals of different frequencies. Then we describe how to use Ohm's Law to analyze capacitive circuits and show how capacitors work closely with resistors to perform useful functions. Finally, we showcase the various uses of capacitors in electronic circuits — proving beyond the shadow of a doubt that capacitors are worth getting charged up about.

Capacitors: Reservoirs for Electrical Energy

When you're thirsty for water, you can generally get a drink in two different ways: by catching water that originated from a source and flows out of pipes when you turn on the tap, or by getting water from a storage container, such as a water cooler. You can think about electrical energy in a similar vein: You can get electrical energy directly from a source (such as a battery or generator), or you can get it from a device that stores electrical energy: a capacitor.

Just as you fill a water cooler by connecting it to a water source, you fill a capacitor with electrical energy by connecting it to a source of electrical energy. And just as water stored in a water cooler remains even after the source is removed, so too the electrical energy stored in a capacitor remains even after the source is removed. In each case, the stuff (water or electrical energy) stored in the device stays there until something comes along and taps into it — whether it's a thirsty consumer or an electronic component in need of electrical energy.

A *capacitor* is a passive electronic device that stores electrical energy transferred from a voltage source. (See Figure 4-1.) If you remove the voltage source and electrically isolate the capacitor (so that it isn't connected in a complete circuit), it holds on to the stored electrical energy. If you connect it to other components in a complete circuit, it will release some or all of the stored energy. A capacitor is made from two metal plates separated by an insulator, which is known as a *dielectric*.

Capacitors and batteries: What's the difference?

Capacitors and batteries both store electrical energy, but in different ways. As we discuss in Chapter 2, a battery uses electrochemical reactions to produce charged particles, which build up on its two metal terminals, creating a voltage. A capacitor doesn't produce charged particles, but it does allow charged particles to build up on its plates, creating a voltage across the plates (see "Charging and Discharging Capacitors"). A battery's electrical energy is the result of an energy conversion process originating from the chemicals stored inside the battery, whereas a capacitor's electrical energy is supplied by a source outside of the capacitor.

Figure 4-1:
Two circuit
symbols
used for
capacitors.

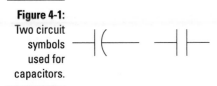

Charging and Discharging Capacitors

If you supply a DC voltage to a circuit containing a capacitor in series with a light bulb (as shown in Figure 4-2), current flow cannot be sustained because there's no complete conductive path across the plates. However, electrons do move around this little circuit — temporarily — in an interesting way.

Figure 4-2:
When a
battery is
placed in a
circuit with
a capaci-
tor, the
capacitor
charges up.
A charged
capaci-
tor stores
electrical
energy,
much like a
battery.

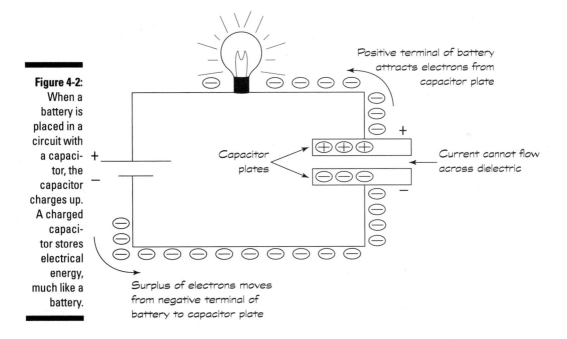

Positive terminal of battery
attracts electrons from
capacitor plate

Capacitor
plates

Current cannot flow
across dielectric

Surplus of electrons moves
from negative terminal of
battery to capacitor plate

Remember that the negative terminal of a battery has a surplus of electrons. So in the circuit shown in Figure 4-2, the surplus electrons begin to move away from the battery toward one side of the capacitor. Once they reach the capacitor, they're stopped in their tracks, with no conductive path to follow across the capacitor. The result is an excess of electrons on that plate.

At the same time, the positive terminal of the battery attracts electrons from the other capacitor plate, so *they* begin to move. As they pass through the light bulb, they light it (but only for a split second, which we explain in the next paragraph). This produces a net positive charge (due to a deficiency of electrons) on that plate. With a net negative charge on one plate, and a net positive charge on the other plate, the result is a voltage difference across the two plates. This voltage difference represents the electrical energy stored in the capacitor.

The battery keeps pushing electrons onto one plate (and pulling electrons off the other plate) until the voltage drop across the capacitor plates is equal to the battery voltage. At this equilibrium point, there is no voltage differential between the battery and the capacitor, so there's no push for electrons to flow from the battery to the capacitor. The capacitor stops charging, and electrons stop moving around through the circuit — and the light bulb goes out.

When the voltage drop across the plates is equal to the battery voltage, the capacitor is said to be *fully charged*. (It's really the capacitor *plates* that are charged; the capacitor as a whole has no net charge.) Even if the battery remains connected, the capacitor will no longer charge up because there is no voltage differential between the battery and the capacitor. If you remove the battery from the circuit, current will not flow and the charge will remain on the capacitor plates. The capacitor looks like a voltage source, as it holds the charge, storing electrical energy.

The larger the battery voltage you apply to a capacitor, the larger the charge that builds up on each plate, and the larger the voltage drop across the capacitor — up to a point. Capacitors have physical limitations: They can handle only so much voltage before the dielectric between the plates is overcome by the amount of electrical energy in the cap, and begins to give up electrons, resulting in current arcing across the plates. You can read more about this in "Keeping an eye on the working voltage," later in this chapter.

If you replace the battery with a simple wire, you provide a path through the bulb for the surplus electrons on one plate to follow to the other (electron-deficient) plate. The capacitor plates *discharge* through the light bulb, lighting it up again briefly — even without a battery in the circuit — until the charge on both plates is neutralized. The electrical energy that had been stored in the capacitor is consumed by the light bulb. When the capacitor is discharged (again, it's really the *plates* that discharge), no more current will flow.

A capacitor can store electrical energy for hours on end. You'd be wise to make sure a capacitor is discharged before handling it, lest it should discharge through you. To discharge a capacitor, carefully place a bulb across its terminals, using insulated alligator clips (discussed in Chapter 9) to make the connection. If the bulb lights up, you know the capacitor was charged, and the light should dim and go out in a few seconds as the capacitor discharges. If you don't have a bulb handy, place a 10kΩ 1 W resistor across the terminals and wait at least 30 seconds.

Opposing voltage change

Because it takes time for charges to build up on the capacitor plates when a DC voltage is applied, and it takes time for the charge to leave the plates once the DC voltage is removed, capacitors are said to "oppose voltage change." This just means that if you suddenly change the voltage applied to a capacitor, it can't react right away; the voltage across the capacitor changes more slowly than the voltage you applied.

Think about being in your car, stopped at a red light. When the light turns green, you get your car moving again, building up speed until you reach the speed limit. It takes time to get to that speed, just like it takes time for a capacitor to get to a certain voltage level. This is quite different from what happens to a resistor. If you switch on a battery across a resistor, the voltage across the resistor changes almost instantaneously.

It takes some time for the capacitor voltage to "catch up" to the source voltage. That isn't a bad thing; in fact, many circuits use capacitors for the express reason that it takes time to charge them up. This is the crux of the reason why capacitors can change the shape (pattern) of electrical signals.

Giving alternating current a pass

Although capacitors cannot pass direct current (DC) (except very briefly, as you saw in the preceding section) because of the dielectric providing a barrier to electron flow, they can pass alternating current (AC).

Suppose you apply an AC voltage source across a capacitor. Remember that an AC voltage source varies up and down, rising from 0 (zero) volts to its peak voltage, then falling back through 0 volts and down to its negative peak voltage, then rising back up through 0 volts to its peak voltage, and so on. Imagine being an atom on one of the capacitor plates and looking at the source terminal nearest you. You'll notice that sometimes you feel a force pulling your electrons away from you, and other times you'll feel a force pushing more electrons toward you. In each case, the strength of the force will vary over time. You and the other atoms on the capacitor plate will alternate between giving up electrons and getting electrons as the source voltage swings up and down.

What's really happening is that as the source voltage rises from 0 volts to its peak voltage, the capacitor charges up, just as it does when you apply a DC voltage. When the supply voltage is at its peak, the capacitor may or may not be fully charged (it depends on a bunch of factors, such as the size of the capacitor plates). Then the source voltage starts to decrease from its peak down to 0 volts. As it does, at some point, the source voltage becomes lower than the capacitor voltage. When this happens, the capacitor starts to discharge through the AC source. Then the source voltage reverses polarity and the capacitor discharges all the way. As the source voltage keeps heading down toward its negative peak voltage, charges start to build up *in reverse* on the capacitor plates: The plate that previously held more negative charges now holds positive charges, and the plate that previously held more positive charges now holds more negative charges. As the source voltage rises from its negative peak, the capacitor again discharges through the AC source, but in the direction opposite to that of its original discharge, and the cycle repeats itself. This continuous charge/discharge cycle can occur thousands — even millions — of times per second, as the capacitor tries to "keep up," so to speak, with the ups and downs of the AC source.

Because the AC source is constantly changing direction, the capacitor goes through a continuous cycle of charging, discharging, and recharging. As a result, electrical charges move back and forth through the circuit, and even though virtually no current flows across the dielectric (except a small *leakage current*), the effect is the same as if current is flowing through the capacitor. So, these amazing capacitors are said to "pass" alternating current (AC) even though they block direct current (DC).

If you add a light bulb to your capacitor circuit powered by an AC voltage source, the bulb will light up and *will stay lit* as long as the AC source is connected. Current alternates its direction through the bulb, but the bulb doesn't care which way current flows through it. (Not so for an LED, or light-emitting diode, which cares very much which way current flows.) Although no current ever actually passes *through* the capacitor, the charging/discharging action of the capacitor creates the effect of current flowing back and forth through the circuit.

What Are Capacitors Used For?

Capacitors are put to good use in most electronic circuits you encounter every day. The key capabilities of capacitors — such as storing electrical energy, blocking DC current, and varying opposition to current depending on applied frequency — are commonly exploited by circuit designers to set the stage for extremely useful functionality in electronic circuits. Here are some of the ways capacitors are used in circuits:

✔ **Storing electrical energy:** Many devices use capacitors to store energy temporarily for later use. Uninterruptible power supplies (UPSs) and alarm clocks keep charged capacitors on hand just in case of a power failure. The energy stored in the capacitor is released the moment the charging circuit is disconnected (which it will be if the power goes out!). Cameras use capacitors for temporary storage of the energy used to generate the flash, and many electronic devices use capacitors to supply energy while the batteries are being changed. Car audio systems commonly use capacitors to supply energy when the amplifier needs more than the car's electrical system can give. Without a capacitor in your system, every time you hear a heavy bass note, your lights will dim!

✔ **Preventing DC current from passing between circuit stages:** When connected in series with a signal source (such as a microphone), capacitors block DC current but pass AC current. This is known as *capacitive coupling* or *AC coupling*, and when used this way, the capacitor is known as a *coupling capacitor*. Multistage audio systems commonly use this functionality between stages so that only the AC portion of the audio signal — the part that carries the encoded sound information — from one stage is passed to the next stage. Any DC current used to power components in a previous stage is removed before the audio signal is amplified.

✔ **Smoothing out voltage:** Power supplies that convert AC to DC often take advantage of the fact that capacitors don't react quickly to sudden changes in voltage. These devices use large electrolytic capacitors to smooth out varying DC supplies. These *smoothing capacitors* keep the output voltage at a relatively constant level by discharging through the load when the DC supply falls below a certain level. This is a classic example of using a capacitor to store electrical energy until you need it: When the DC supply can't maintain the voltage, the capacitor gives up some of its stored energy to take up the slack.

✔ **Creating timers:** Because it takes time to charge and discharge a capacitor, capacitors are often used in timing circuits to create "ticks" and "tocks" when the voltage rises above or falls below a certain level. The timing of the ticking and tocking can be controlled through selection of the capacitor and other circuit components. (For details, see the section called "Teaming Up with Resistors.")

✔ **Tuning in (or out) frequencies:** Capacitors are often used to help select or reject certain electrical signals, depending upon their frequency. For instance, a tuning circuit in a radio-receiver system relies on capacitors and other components to allow the signal from just one radio station at a time to pass through to the amplifier stage, while blocking signals from all other radio stations. Each radio station is assigned a specific broadcast frequency, and it's the radio builder's job to design circuits that "tune in" target frequencies. Because capacitors behave differently for different signal frequencies, they are a key component of these tuning circuits. The net effect is a kind of electronic filtering. (For more about simple electronic filters, see the section called "Selecting Frequencies with Simple RC Filters.")

Characterizing Capacitors

There are lots of ways to build capacitors, using different materials for the plates and dielectric and varying the size of the plates. The particular makeup of a capacitor determines its characteristics and influences its behavior in a circuit.

How much charge can a capacitor plate store?

Capacitance is the ability of a body to store charge. The same term — capacitance — is used to describe just how much charge a capacitor can store on either one of its plates. The higher the capacitance, the more charge the capacitor can store at any one time.

The capacitance of any given capacitor depends on three things: the surface area of the metal plates, the thickness of the dielectric between the plates, and the type of dielectric used (more about dielectrics later in this section).

You don't need to know how to calculate capacitance (and, yes, there is a scary-looking formula), because any capacitor worth its salt will come with a documented capacitance value. It just helps to understand that how much charge a capacitor's plates can hold depends on how the capacitor is made.

Capacitance is measured in units called *farads.* One farad (abbreviated F) is defined as the capacitance needed to get one amp of current to flow when the voltage changes at a rate of one volt per second. Don't worry about the details of the definition; just know that one farad is a very, very, very large amount of capacitance. You're more likely to run across capacitors with much smaller capacitance values — hovering in the microfarad (µF) or pico-farad (pF) range. A microfarad is a millionth of a farad, or 0.000001 farad, and a picofarad is a millionth of a millionth of a farad, or 0.000000000001 farad.

Here are some examples:

- ✔ A 10 µF capacitor is 10 millionths of a farad.
- ✔ A 1 µF capacitor is 1 millionth of a farad.
- ✔ A 100 pF capacitor is 100 millionths of a millionth of a farad, or you could say it is 100 millionths of a microfarad. Whew!

You'll find larger capacitors (1F or more) used for system energy storage, while smaller capacitors are used for a variety of applications, as shown in Table 4-1.

Most capacitors are rather inexact beasts. The actual capacitance of the capacitor can vary quite a bit from the nominal capacitance. Manufacturing variations cause this problem; capacitor makers aren't just out to confuse you. Fortunately, the inexactness is seldom an issue in homebrewed circuits. Still, you need to know about these variations so that if a circuit calls for a higher-precision capacitor, you know what to buy. As with resistors, capacitors are rated by their tolerance, which is expressed as a percentage.

Big capacitor in itty-bitty living space

Making farad-range capacitors has become possible only recently. Using older construction techniques, a one-farad capacitor would be bigger than a bread box and kind of unwieldy.

By using other technologies and materials, such as microscopically small carbon granules, manufacturers can now build capacitors of one farad and above that fit into the palm of your hand. Computer memories, clock radios, and other electric devices that need to retain a small charge for extended periods of time when they have no access to power routinely use capacitors as substitute batteries.

Keeping an eye on the working voltage

The *working voltage,* sometimes abbreviated as *WV,* is the highest voltage that the manufacturer recommends placing across a capacitor safely. If you exceed the working voltage, you may damage the dielectric, which could result in current arcing between the plates, like a lightning strike during a storm. That means you can short out your capacitor and allow all sorts of unwanted current to flow — and maybe even damage nearby components.

Capacitors designed for DC circuits are typically rated for a WV of no more than 16 V to 35 V. That's plenty for DC circuits, which are usually powered by sources ranges from 3.3 V to 12 V. If you build circuits that use higher voltages, be sure to select a capacitor that has a WV of at least 10 percent to 15 percent more than the supply voltage in your circuit, just to be on the safe side.

Choosing the right type (dielectric) for the job

Designers of electronic circuitry specify capacitors for projects by the dielectric material in them. Some materials are better in certain applications and are inappropriate for other applications. For instance, ceramic capacitors perform reliably only for signal frequencies of less than 100,000 hertz, whereas mica capacitors exhibit exceptional frequency characteristics and are often used in precision timing and filtering circuits.

The most common dielectric materials are aluminum electrolytic, tantalum electrolytic, ceramic, mica, polypropylene, polyester (or Mylar), and polystyrene. If a circuit diagram calls for a capacitor of a certain type, be sure you get one that matches.

Table 4-1 lists the most common capacitor types, their typical value range, and common applications.

Table 4-1	Capacitor Characteristics	
Type	*Typical Range*	*Application*
Ceramic	1 pF to 2.2 µF	Filtering, bypass
Mica	1 pF to 1 µF	Timing, oscillator, precision circuits
Metalized foil	to 100 µF	DC blocking, power supply, filtering
Polyester (Mylar)	0.001 to 100 µF	Coupling, bypass
Polypropylene	100 pF to 50 µF	Switching power supply
Polystyrene	10 pF to 10 µF	Timing, tuning circuits
Tantalum (electrolytic)	0.001 to 1,000 µF	Bypass, coupling, DC blocking
Aluminum electrolytic	10 to 220,000 µF	Filtering, coupling, bypass, smoothing

Sizing up capacitor packaging

Capacitors come in a variety of shapes and sizes, as shown in Figure 4-3. Aluminum electrolytic and paper capacitors commonly come in a cylindrical shape. Tantalum electrolytic, ceramic, mica, and polystyrene capacitors have a more bulbous shape because they typically get dipped into an epoxy or plastic bath to form their outside skin. However, not all capacitors of any particular type (such as mica or Mylar) get manufactured the same way, so you can't always tell the component book by its cover.

Your favorite parts supplier may label capacitors according to the way their leads are arranged: axially or radially. Axial leads extend out from either end of a cylindrically shaped capacitor, along its axis; radial leads extend from one end of a capacitor and are parallel to each other (until you bend them for use in a circuit).

If you go searching for capacitors inside your personal computer (PC), you may not recognize some of them when you see them. That's because many of the capacitors in your PC don't have any leads at all! So-called *surface-mount packages* for capacitors are extremely small and are designed to be soldered directly to printed circuit boards (PCBs) such as the ones in your PC. Since the 1980s, high-volume manufacturing processes have been using surface-mount technology (SMT) to connect capacitors and other components directly to the surface of PCBs, saving space and improving circuit performance.

Being positive about capacitor polarity

Some larger-value electrolytic capacitors (1µF and up) are *polarized* — meaning that the positive terminal must be kept at a higher voltage than the negative terminal, so it matters which way you insert the capacitor into your circuit. Polarized capacitors are designed for use in DC circuits.

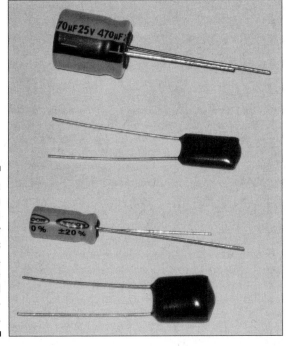

Figure 4-3:
Capacitors come in a variety of shapes and sizes, and may be polarized or non-polarized.

Many polarized capacitors sport a minus (−) sign or a large arrow pointing toward the negative terminal. For radial capacitors, the negative lead is often shorter than the positive lead.

Just because capacitor packages generally point out the negative terminal doesn't mean circuit diagrams follow the same convention. Usually, if there's a polarized capacitor in a circuit, you will see a plus sign (+) on one side of the capacitor symbol, shown in Figure 4-4, showing you how to orient the capacitor in the circuit.

If a capacitor is polarized, you *really, really* need to make sure to install it in the circuit with the proper orientation. If you reverse the leads, say, by connecting the + side to the ground rail in your circuit, you may cause the dielectric inside the capacitor to break down, which could effectively short-circuit the capacitor. This may damage other components in your circuit (by sending too much current their way), and your capacitor may even explode.

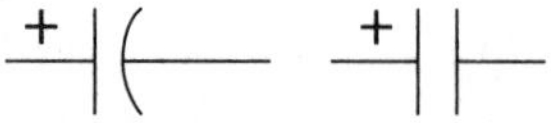

Figure 4-4:
Symbols for
polarized
capacitors.

Reading into capacitor values

Some capacitors have their values printed directly on them, either in farads or portions of a farad. You usually find this to be the case with larger capacitors, because there is enough real estate for printing the capacitance and working voltage.

Most smaller capacitors (such as 0.1 µF or 0.01 µF mica disc capacitors) use a three-digit marking system to indicate capacitance. Most folks find the numbering system easy to use. But there's a catch! (There's always a catch.) The system is based on *pico*farads, not microfarads. A number using this marking system, such as 103, means 10, followed by three zeros, as in 10,000, for a grand total of 10,000 picofarads. Some capacitors have a two-digit number printed on them, which is simply its value in picofarads. For instance, a value of 22 means 22 picofarads. No third digit means no zeros to tag on to the end.

For values over 1,000 picofarads, your parts supplier will most likely list the capacitor in microfarads, even if the markings on it indicate picofarads. To convert the picofarad value you read on the capacitor into microfarads, just move the decimal point six places to the *left*. So a capacitor marked with a 103 (say, the example in the preceding paragraph) has a value of 10,000 pF or 0.01 µF.

Suppose you're building a circuit that calls for a 0.1 µF disc capacitor. You can convert microfarads into picofarads to figure out what marking to look for on the capacitor package. Just move the decimal point six places to the *right*, and you get 100,000 pF. Because the three-digit marking consists of the first two digits of your pF value (10) followed by the additional number of zeros (4), you'll need a mica disc capacitor labeled "104."

You can use Table 4-2 as a reference guide to common capacitor markings that use this numbering system.

Table 4-2	Capacitor Value Reference
Marking	*Value*
nn (a number from 01 to 99)	nn pF
101	100 pF

Marking	Value
102	0.001 µF
103	0.01 µF
104	0.1 µF
221	220 pF
222	0.0022 µF
223	0.022 µF
224	0.22 µF
331	330 pF
332	0.0033 µF
333	0.033 µF
334	0.33 µF
471	470 pF
472	0.0047 µF
473	0.047 µF
474	0.47 µF

Another, less-often-used numbering system uses both numbers and letters, like this: 4R1

The placement of the letter R tells you the position of the decimal point: 4R1 is really 4.1. This numbering system doesn't indicate the units of measure, however, which can be either microfarads or picofarads.

You can test capacitance with a capacitor meter or a multimeter with a capacitance input. Most meters require that you plug the capacitor directly into the test instrument, as the capacitance can increase with longer leads. This makes the reading less accurate.

On many capacitors, a single-letter code indicates the tolerance. You may find that letter placed by itself on the body of the capacitor or placed after the three-digit mark, like this: 103Z.

Here the letter Z denotes a tolerance of +80 percent to –20 percent. This tolerance means that the capacitor, rated at 0.01 µF, may have an actual value that's as much as 80 percent higher — or 20 percent lower — than the stated value. Table 4-3 lists the meanings of common code letters used to indicate capacitor tolerance. Notice that the letters B, C, and D represent tolerances expressed in absolute capacitance values, rather than percentages. These markings are used only on very small (pF range) capacitors.

Table 4-3	Capacitor Tolerance Markings
Code	*Tolerance*
B	± 0.1 pF
C	± 0.25 pF
D	± 0.5 pF
F	± 1%
G	± 2%
J	± 5%
K	± 10%
M	± 20%
P	+100%, −0%
Z	+80%, −20%

Varying capacitance

Variable capacitors allow you to adjust capacitance to suit your circuit's needs. The symbols for a variable capacitor are shown in Figure 4-5.

Figure 4-5:
Symbols
that stand
for variable
capacitors.

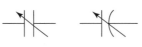

The most common type of variable capacitor is the *air dielectric,* a type found frequently in the tuning controls of AM radios. Smaller-variable capacitors are often used in radio receivers and transmitters, and they work in circuits that use quartz crystals to provide an accurate reference signal. The value of such variable capacitors typically falls in the 5- to 500-pF range.

Mechanically controlled variable capacitors work by changing the distance between the capacitor plates or by changing the amount of overlap between the surfaces of the two plates. A specially designed *diode* (a semiconductor device discussed in Chapter 6) can function as an electronically controlled variable capacitor; such devices are known as *varactors* or *varicaps* — and you can change their capacitance by changing the DC voltage you applied to them.

Chances are, you interact with variable capacitors more than you do with your spouse. Variable capacitors are what's behind many touch-sensitive devices, such as the keys on computer keyboards, control panels on many appliances and in some elevators, and the buttons on your favorite remote control. One type of microphone uses a variable capacitor to convert sound into electrical signals, with the diaphragm of the mike acting as a movable capacitor plate. Sound fluctuations make the diaphragm vibrate, which varies the capacitance, producing voltage fluctuations. This device is known as a condenser microphone, so named because capacitors used to be called condensers.

Combining Capacitors

If you want to analyze multiple capacitors in a circuit, you can combine them to get the equivalent capacitance. But as you'll see, the rules for combining capacitors are different from the rules for combining resistors.

Capacitors in parallel

Figure 4-6 shows two capacitors in parallel, with the common connection points labeled A and B. Notice that point A is connected to one plate of capacitor C1 and one plate of capacitor C2. Electrically speaking, point A is connected to a metal plate that is the size of the two plates combined. Likewise for point B, which is connected to the both the other plate of capacitor C1 and the other plate of capacitor C2. The larger the surface area of a capacitor plate, the larger the capacitance.

Capacitors in parallel add up: Each metal plate of one capacitor is tied electrically to one metal plate of the parallel capacitor. Each pair of plates behaves as a single larger plate with a higher capacitance, as shown in Figure 4-6.

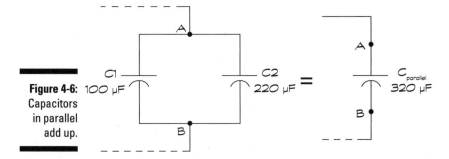

Figure 4-6: Capacitors in parallel add up.

The equivalent capacitance of a set of capacitors in parallel is

$$C_{parallel} = C1 + C2 + C3 + C4 + \ldots$$

C1, C2, C3, and so forth represent the values of the capacitors and $C_{parallel}$ represents the total equivalent capacitance.

For the capacitors in Figure 4-6, the total capacitance is

$$C_{parallel} = 100 \ \mu F + 220 \ \mu F$$
$$= 320 \ \mu F$$

If you place the capacitors from Figure 4-6 in a working circuit, the voltage across each capacitor will be the same, and current flowing in to point A will split up to travel through each capacitor, and then sum together again at point B.

Capacitors in series

Capacitors placed in series work against each other, reducing the effective capacitance in the same way that resistors in parallel reduce the overall resistance. The calculation looks like this:

$$C_{series} = \frac{C1 \times C2}{C1 + C2}$$

C1 and C2 are the values of the individual capacitors, and C_{series} is the equivalent capacitance. The total capacitance (in μF) of a 100-μF capacitor in series with a 220-μF capacitor, as shown in Figure 4-7, is

$$C_{series} = \frac{100 \times 220}{100 + 220}$$
$$= \frac{22,000}{320}$$
$$= 68.75$$

$$C_{series} = 68.75 \ \mu F$$

You can temporarily ignore the "μ" part of "μF" while you're performing the calculation just shown — as long as all the capacitance values are in μF and you remember that the resulting *total* capacitance is also in μF.

The equivalent capacitance of a set of capacitors in series is:

$$C_{series} = \frac{1}{\frac{1}{C1} + \frac{1}{C2} + \frac{1}{C3} \ldots \text{(and more as needed)}}$$

As for any components in series, the current running through each capacitor in series is the same, but the voltage dropped across each capacitor may be different.

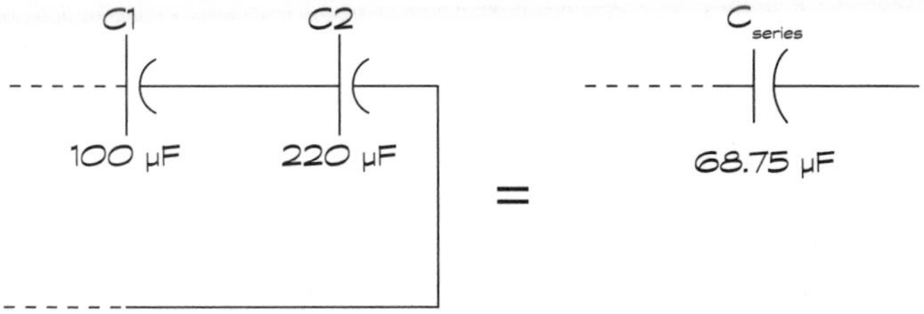

Figure 4-7:
Capacitors
in series
work
against
each other,
reducing
the overall
capacitance.

C1

C2

C_{series}

100 µF

220 µF

68.75 µF

=

Understanding Capacitive Reactance

In Chapter 3, we define resistance as a measure of an object's opposition to the flow of electrons, and we said that resistors have controlled amounts of resistance that remains the same no matter what the voltage or current situation. If you could measure a capacitor's opposition to the flow of electrons, however, you would find that it varies, depending on the situation.

Earlier in this chapter, we said that capacitors block direct current (except for a short time while they're charging and discharging) and allow alternating current. When a DC voltage is suddenly applied to a capacitor, as in the light-bulb circuit shown in Figure 4-2, here's what happens:

1. At first, while the capacitor is charging, current flows freely through the circuit, lighting the bulb.

2. As the capacitor nears its capacity to hold charge, less current flows until eventually the capacitor is fully charged, and current no longer flows.

The capacitor provides very little opposition to electron flow when a sudden voltage change is first applied — but eventually it behaves like an open circuit, opposing all electron flow. When an AC voltage source is applied to a capacitive circuit, current is allowed to flow through the circuit. The faster the source voltage fluctuates, the less the capacitor opposes electron flow — just like when the battery was suddenly applied to the discharged capacitor in Figure 4-2. The slower the source voltage fluctuates, the more the capacitor opposes electron flow.

The apparent AC resistance to current is known as the *capacitive reactance,* and is measured in ohms (yes, ohms!). Capacitive reactance is similar to resistance in that it represents opposition to current. However, unlike resistance, which is constant for a particular resistive device, capacitive reactance varies depending upon the frequency of the voltage applied to the capacitor.

You calculate capacitive reactance, symbolized by X_c, using the following formula:

$$X_c = (2 \times \pi \times f \times C)^{-1}$$

In this formula, f represents the frequency in hertz (Hz) of the applied AC voltage, C is the capacitance in farads (*not* μF or pF), and Π is that constant you first met in geometry class that starts out as 3.14 and keeps trailing numbers off to the right of the decimal point without repeating itself. You can approximate $2 \times \Pi$ as 6.28 and simplify the formula as follows:

$$X_c \approx (6.28 \times f \times C)^{-1}$$

You can see from this formula that capacitive reactance *decreases* as the frequency of the applied voltage *increases*. (A higher value of frequency makes the denominator bigger, which makes the overall fraction smaller.) For instance, you can calculate the capacitive reactance of a 0.1-μF capacitor when an AC voltage source fluctuating at 20 kHz is applied as follows:

$$X_c \approx (6.28 \times 20,000 \times 0.0000001)^{-1}$$
$$\approx 80 \text{ ohms}$$

If you slow down the frequency of the source voltage to 1 Hz, the capacitive reactance changes like this:

$$X_c \approx (6.28 \times 1 \times 0.0000001)^{-1}$$
$$\approx 1.6 \text{ Mohms}$$

Note that this is *very* different from the constant-no-matter-what resistance of your average resistor. For capacitors, the faster the voltage fluctuates (that is, the higher the frequency of its fluctuation), the lower the reactance and the more freely current will flow. The slower the voltage fluctuates (that is, the lower the frequency), the higher the reactance — and the less easily current will flow. If the frequency is zero, which means no voltage fluctuations (or a constant DC voltage), the denominator is zero, and the reactance is infinite. That's the open circuit (infinite resistance) situation in which the capacitor blocks DC signals.

Using Ohm's Law for capacitive reactance

Because capacitive reactance is measured in ohms, you might be wondering if you can use Ohm's Law for capacitors. The answer is yes — sort of. Ohm's Law works for capacitive reactance — but just *one frequency at a time*. If you change the frequency of the AC voltage, even if you don't change the magnitude of the voltage ups and downs, you have to recalculate Ohm's Law given the new frequency.

Suppose you apply an AC voltage source with a peak value of 5 volts to your 0.1μF capacitor. The current through the circuit will alternate at the same frequency as the voltage, but Ohm's Law tells you that the peak value of the current depends on the peak value of the voltage and the capacitive reactance. Say (for example) the frequency is 1 Hz — from calculations in the preceding section, you know that the capacitive reactance of a 0.1-μF capacitor at 1 Hz is 1.6 MΩ. Now you can apply Ohm's Law to calculate the peak AC current "through" the capacitor for a 1-Hz signal as follows:

$$I_{peak} = \frac{V_{peak}}{X_c}$$

$$= \frac{5 \text{ volts}}{1,600,000 \text{ ohms}}$$

$$\approx 0.0000031 \text{ amps or } 3.1 \text{ } \mu A$$

Say you change the frequency to 20 kHz, but keep the same peak voltage of 5 volts. Now the capacitive reactance is 80 Ω (as calculated in the preceding section). Using Ohm's Law, the peak current "through" the capacitor when you apply a 20-kHz voltage with a peak value of 5 volts is

$$I_{peak} = \frac{V_{peak}}{X_c}$$

$$= \frac{5 \text{ volts}}{80 \text{ ohms}}$$

$$\approx 0.0625 \text{ amps or } 62.5 \text{ mA}$$

So for the same capacitor in the circuit, as you increase the frequency of the source voltage, you decrease the capacitive reactance, resulting in an increase in current flowing through the circuit. Likewise, if you decrease the frequency of the source voltage, you increase the capacitive reactance, resulting in a decreased current.

Unlike resistors, how capacitors in an AC circuit behave depends on the frequency of the voltage applied to them. You can (and should) exploit this frequency-dependent behavior to create circuits that perform useful functions — for example, filters that favor high-frequency signals over low-frequency signals (and vice versa). (See "Selecting Frequencies with Simple RC Filters.")

Teaming Up with Resistors

Capacitors are often found working hand in hand with resistors in electronic circuits, combining their talent for storing electrical energy with a resistor's control of electron flow. Put these two capabilities together and you can control how fast electrons fill (or charge) a capacitor — and how fast those electrons empty out (or discharge) from a capacitor. This dynamic duo is so popular that circuits containing both *resistors* and *capacitors* are known by a handy nickname: *RC circuits.*

Timing is everything

Take a look at the RC circuit in Figure 4-8: The battery will charge the capacitor through the resistor when the switch is closed. At first, the voltage across the capacitor, V_c, is zero (assuming the capacitor was discharged to begin with). When you close the switch, current starts to flow and charges start to build up on the capacitor plates. Ohm's Law tells you that the charging current, I, is determined by the voltage across the resistor, V_r, and the value of the resistor, $R\left(I = \dfrac{V_r}{R}\right)$. And because the voltage drops equal the voltage rises around the circuit, you know that the resistor voltage is the difference between the supply voltage, V_{supply}, and the capacitor voltage, V_c ($V_r = V_{supply} - V_c$). Using those two facts, you can analyze what is going on in this circuit over time, as follows:

Figure 4-8:
In an RC circuit, the capacitor charges up through the resistor. The values of the resistor and capacitor determine how quickly the capacitor charges.

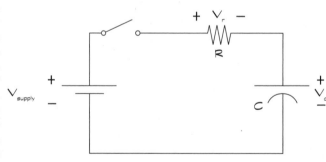

✔ **Initially:** Because the capacitor voltage is initially zero, the resistor voltage is initially equal to the supply voltage.

✔ **Charging:** As the capacitor begins to charge, it develops a voltage, so the resistor voltage begins to fall, which in turn reduces the charging current. The capacitor continues to charge, but at a slower rate because the charging current has decreased. As V_c continues to increase, V_r continues to decrease, so the current continues to decrease.

✔ **Fully charged:** When the capacitor is fully charged, current stops flow-ing, the voltage drop across the resistor is zero, and the voltage drop across the capacitor is equal to the supply voltage.

If you remove the battery, and connect the resistor in parallel with the capac-itor, the capacitor will discharge through the resistor. This time, the voltage across the resistor is equal to the voltage across the capacitor ($V_r = V_c$), so the current is V_c/R. Here's what happens:

✔ **Initially:** Because the capacitor is fully charged, its voltage is initially V_{supply}. Because $V_r = V_c$, the resistor voltage is initially V_{supply}, so the current jumps up immediately to $\dfrac{V_{supply}}{R}$. This means the capacitor is shuffling charges from one plate to the other pretty quickly.

✔ **Charging:** As charges begin to flow from one capacitor plate to the other, the capacitor voltage (and so V_r) starts to drop, resulting in a lower current. The capacitor continues to discharge, but at a slower rate. As V_c (and so V_r) continues to decrease, so does the current.

✔ **Fully discharged:** When the capacitor is fully discharged, current stops flowing, and no voltage is dropped across either the resistor or the capacitor.

The waveform in Figure 4-9 shows how, when a constant voltage is applied and then removed from the circuit, the capacitor voltage changes over time as it charges and discharges through the resistor. How fast the capacitor charges (and discharges) depends on the resistance and capacitance of the RC circuit. The larger the resistance, the smaller the current that flows for the same supply voltage — and the longer it takes the capacitor to charge. A smaller resistance allows more current to flow, charging the capacitor faster. Likewise, the larger the capacitance, the more charges it takes to fill the capacitor plates, so the longer it takes to charge the capacitor. During the discharge cycle, a larger resistor slows down the electrons more as they move from one plate to the other, increasing the discharge time, and a larger capacitor holds more charge, taking longer to discharge.

Figure 4-9:
The voltage across a capacitor changes over time as the capaci-tor charges and dis-charges.

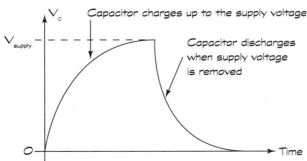

Calculating RC time constants

By picking the values of the capacitor and the resistor carefully, you can adjust a capacitor's charge and discharge time. As it turns out, your choice of resistance, R, and capacitance, C, *defines* the time it takes to charge and discharge your chosen capacitor through your chosen resistor. If you multiply R (in ohms) by C (in farads), you get what is known as the *RC time constant* of your RC circuit, symbolized by T. And that makes another handy formula:

$$T = R \times C$$

A capacitor charges and discharges almost completely after five times its RC time constant, or 5RC (which really means $5 \times R \times C$). After time has passed that's equivalent to one time constant, a discharged capacitor will charge to roughly two-thirds its capacity — and a charged capacitor will discharge nearly two-thirds of the way.

For instance, suppose you choose a 2-MΩ resistor and a 15-μF capacitor for the circuit in Figure 4-8. You calculate the RC time constant as follows:

RC time constant = $R \times C$ = 2,000,000 ohms $\times$ 0.000015 farad = 30 seconds

Then you know that it will take about 150 seconds (or $2^1/_2$ minutes) to fully charge or discharge the capacitor. If you'd like a shorter charge/discharge cycle time, you can reduce the value you choose for either the resistor or the capacitor (or both). Suppose that 15-μF capacitor is the only one you have in the house, and you want to charge it in five seconds. You can figure out what resistor you need to make this happen as follows:

✔ **Find the RC time constant:** You know that it takes five times the RC time constant to fully charge the capacitor, and you want to fully charge your capacitor in five seconds. That means that 5RC = 5 seconds, so RC = 1 second.

✔ **Calculate R:** If $R \times C$ = 1 second, and C is 15-μF, then you know that $R = {}^{1\ second}/_{0.000015\ farad}$, which is approximately 66,667 ohms or 67 kΩ.

Creating a timer

Armed with the knowledge of the RC time constant, you can use an RC circuit to create a timer. Say, for example, your freeloading cousin is visiting for a month or so, and he keeps raiding your refrigerator in the middle of the night. You decide to set up a little noisemaker to scare him off whenever he opens the fridge. Just for fun, you'd like to allow him some time to gaze at the succulent food inside before you shock the pants off him with a few choice decibels of sound from a buzzer, triggered by a switch that closes when he opens the fridge door.

If you've got a buzzer that requires a voltage of 6 volts in order to sound off, and you're using a 9-volt battery to power your little scare circuit, you can build an RC circuit like the one in Figure 4-8 and use the capacitor voltage to trigger the buzzer. The idea is to charge the capacitor to about 6 volts in the time you'd like your cousin to salivate over your food — and then blast the buzzer.

Suppose you want your cousin to enjoy the view for about 10 seconds. You've got a 15-µF capacitor handy, so you need to calculate the resistance needed to charge the capacitor to 6 volts in 10 seconds. Because the capacitor will keep charging until it reaches the full-power supply voltage of 9 volts, the trigger point (when the buzzer will sound) is when the capacitor reaches $^6/_9$ of its capacity, or about two-thirds capacity. That happens after just one time constant. You calculate the resistance for your 10-second delay-to-buzzing interval as follows:

10 seconds = R × 0.000015 farad

$R = {}^{10\text{ seconds}}/_{0.000015\text{ farad}} \approx 667$ ohms

You have a 620 Ω resistor handy, so your actual RC time constant is about 9.3 seconds (that's 620 ohms × 15-µF). It will take about 9.3 seconds for the capacitor voltage to reach 6 volts, so your cousin will have enough time to look — but not enough time to eat — before he gets caught red-handed.

If you want to fine-tune the delay, use a resistor with a slightly smaller value than you need and add a potentiometer (pot) in series with the resistor. Because the total resistance is the sum of the value of the fixed resistor and that of the pot, you can increase or decrease the resistance by adjusting the pot. Just tweak the pot until you get the delay you want. (Chapter 3 covers potentiometers in more detail.)

Selecting Frequencies with Simple RC Filters

Because capacitors behave differently depending on the frequency of the voltage or current in a circuit, they're often used in special circuits called *filters* to allow or reject various signals. Capacitors naturally block DC signals and allow AC signals to pass, but you can control just which AC signals are allowed to pass by carefully choosing the components in your filter circuits.

This section looks at very simple filters — and at how you can control which frequencies make it through them. Electronic filter design, which is a field of study on its own and beyond the scope of this book, usually involves creating more complex circuits for precise control of the output. The underlying concepts, however, are the same as for the simple filters described here.

Low-pass filter

In the circuit in Figure 4-10, a variable voltage source, labeled V_{in}, is applied to a series RC circuit, and the output of the circuit, V_{out}, is the voltage across the capacitor. Suppose you apply a constant voltage (f = 0 Hz) to this circuit. No current flows, so the full input voltage is dropped across the capacitor: V_{out} = V_{in}. At other end of the frequency spectrum, the capacitive reactance for a very, very high frequency is a very, very low value. This effectively short-circuits the capacitor, so the voltage drop across the capacitor is zero: V_{out} = 0.

As you alter the frequency of the input signal from very low to very high, the capacitive reactance varies from very high to very low. The higher the reactance, the more voltage is dropped across the capacitor (at the expense of voltage dropped across the resistor). The lower the reactance, the less voltage is dropped across the capacitor (and the more is dropped across the resistor). This circuit tends to allow lower frequencies to pass from input to output, while blocking higher frequencies from getting through — so it's known as a *low-pass filter.*

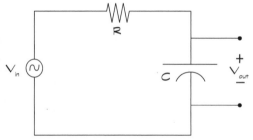

Figure 4-10: A low-pass filter allows input signals with low frequencies to pass through to the output.

High-pass filter

By reversing the roles of the resistor and capacitor in the low-pass RC circuit, you can create the opposite effect: a *high-pass filter.* In the circuit in Figure 4-11, the output voltage is the voltage across the resistor. For very-low-frequency input signals, the capacitor blocks current from flowing, so no voltage is dropped across the resistor: V_{out} = 0. For very-high-frequency input signals, the capacitor acts like a short circuit, so current flows and all the input voltage is dropped across the resistor: V_{out} = V_{in}.

As you alter the frequency from low to high, the capacitive reactance varies from high to low. You can think of this as placing an imaginary device — a frequency-controlled potentiometer — in the circuit in place of the capacitor: As the input frequency increases, the reactance decreases — and the more input voltage is dropped across the resistor.

Figure 4-11:
A high-pass
filter allows
input signals
with high
frequen-
cies to pass
through to
the output.

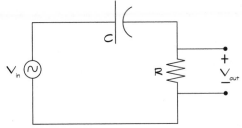

Cutting off frequencies at the knees

Filter circuits are designed to pass certain frequencies while *attenuating*,
or reducing the amplitude of, other frequencies. No filter is perfect; it can't
pass all signals above or below a specific frequency completely, while totally
blocking all other frequencies. Complex filter designs are much better than
simple RC filters at distinguishing which frequencies get through and which
do not, but all filters — whether simple or complex — share a design parameter known as the *cut-off frequency.*

The *cut-off frequency,* F_c, is the frequency at which the filter starts to restrict
the passage of the input signal. Figure 4-12 shows a graph of the amplitude of
the output signal for various input frequencies for a high-pass filter (notice
that this graph plots frequency, not time!). The graph shows that frequencies
above the cut-off value are allowed to pass with little or no attenuation, while
frequencies below the cut-off value are attenuated substantially. The cut-off
frequency occurs near the knee of the curve; you can find it by using this
equation:

$$F_c = (2 \times \pi \times T)^{-1}$$

Because $T = R \times C$, you can control the cut-off frequency of your simple low-
or high-pass filter by carefully selecting the values of the resistor and capacitor, according to the following equation:

$$F_c = (2 \times \pi \times R \times C)^{-1}$$

For instance, say you have a high-pass filter configuration with a 220Ω resistor and a 0.1-μF capacitor. The cut-off frequency of your filter will be roughly
$1/6.28 \times 220$ ohms $\times 0.0000001$ farad, or about 7,200 Hz. If you use such a filter
in an audio system, don't be surprised if you don't hear much of your favorite band's voices or their instruments: The sounds they make fall well below
7,000 Hz, and your simple filter attenuates those sounds!

Figure 4-12:
The filter
cut-off
frequency,
F_c, is the
frequency
at which the
filter starts
to attenuate
the signal.

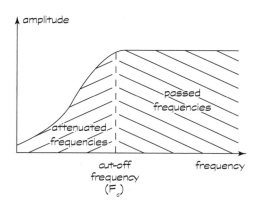

Filtering frequency bands

You can design filters to have two cut-off frequencies — one high and one low — to either allow a certain *band,* or range, of frequencies to pass through or reject a band of frequencies. Such filters are known as *band-pass* and *band-reject* filters, respectively, and are constructed by combining low- and high-pass filters in just the right way.

Band-pass filters are commonly used in radio-receiver systems to select just the right signal from among many signals transmitted through the air. You might use a band-reject filter to "filter out" unwanted hum from a 60 Hz power line — as long as you know the frequency range of the hum. Most of these complex filters employ inductors as well as capacitors and resistors. We discuss inductors in Chapter 5.

Trying Your Hand at Simple Capacitive Circuits

You can peek ahead to Chapter 14 in Part III if you're anxious to play with some real capacitive circuits. The learning circuits in that chapter will get you charged up over the filling and emptying of a capacitor, and give you first-hand experience with the magic of the RC time constant. If you do decide to charge ahead, we ask you to please read through Part II first, so that you know how to set up your electronics shop, read schematics, and — most importantly — practice safe circuit-building.

Chapter 5

Curling Up with Coils and Crystals

. . . .

In This Chapter

▶ Inducing currents in coils with a changing magnetic field

▶ Opposing changes in current with an inductor

▶ Using inductors in filter circuits

▶ Resonating with RLC circuits

▶ Making frequencies crystal-clear

▶ Coupling magnetic flux to transfer energy between circuits

. . . .

Many of the best inventions in the world, including penicillin, sticky notes, champagne, and the pacemaker, were the result of pure accidental discovery (in some cases, attributable to utter carelessness or second-rate science). One such serendipitous discovery — the interaction between electricity and magnetism — led to the development of two amazingly useful electronic components: the induction coil and the transformer.

The *induction coil,* or *inductor,* stores electrical energy in a magnetic field and shapes electrical signals in a different way than does a capacitor. Whether operating alone, in special pairs known as *transformers,* or as part of a team along with capacitors and resistors, inductors are at the heart of many modern-day conveniences you might not want to live without, including radio systems, television, and — oh, a minor thing — the electric power-transmission network.

This chapter exposes the relationship between electricity and magnetism and explains how 19th-century scientists purposely exploited that relationship to create inductors and transformers. You get a look at what happens when you try to change the direction of current through an inductor too quickly, and how Ohm's Law can be applied to inductors. Then you explore how inductors are used in circuits and why crystals ring at just one frequency. Finally, you get a (well-insulated) handle on how transformers transfer electrical energy from one circuit to another — without any direct contact between the circuits.

Kissing Cousins: Magnetism and Electricity

Magnetism and electricity were once thought to be two completely separate phenomena, until a 19th-century scientist named Hans Christian Ørsted discovered that a compass needle moved away from magnetic north when current supplied by a nearby battery was switched on and off. Ørsted's keen observation led to lots of research and experimentation, ultimately confirming the fact that electricity and magnetism are closely related. After several years (and many more accidental discoveries), Michael Faraday and other 19th-century scientists figured out how to capitalize on the phenomenon known as *electromagnetism* to create the world's first electromechanical devices. Today's power transformers, electromagnetic generators, and many industrial motors are based on the principles of electromagnetism.

This section looks at how electricity and magnetism interact.

Drawing the (flux) lines with magnets

Just as electricity involves a force (voltage) between two electrical charges, so magnetism involves a force between two magnetic poles. If you've ever performed the classic grade-school science experiment where you place a magnet on a surface and toss a bunch of iron filings near the magnet, you've seen the effects of magnetic force. Remember what happened to the filings? They settled into curved linear paths from the north pole of the magnet to its south pole. Those filings showed you the magnetic lines of force — also known as *flux lines* — within the magnetic field created by the magnet. You may have seen more filings closer to the magnet because that's where the magnetic field is strongest. Figure 5-1 shows the pattern produced by invisible lines of flux around a magnet.

Magnetic flux is just a way to represent the strength and direction of a magnetic field. To understand magnetic lines of flux, think about the effects of air on a sailboat's sail. The stronger the wind and the larger the sail, the greater the force of air on the sail. But if the sail is oriented parallel to the direction of wind flow, air slips by the sail and even a strong wind will not move the sail. The effect of the wind is greatest when it hits the sail head-on — that is, when the surface of the sail is perpendicular to the direction of wind flow. If you try to represent the strength and direction of the wind and the orientation of the sail in a diagram, you might draw arrows showing the force of the wind extending through the surface of the sail. Likewise, lines of magnetic flux illustrate the strength and orientation of a magnetic field, showing you how the force of the magnetic field will act on an object placed within the field. Objects placed in the magnetic field will be maximally affected by the force of the magnetic field if they are oriented perpendicular to the flux lines.

Figure 5-1:
Magnetic
lines of
force exist
in parallel
flux lines
from a mag-
net's north
pole to its
south pole.

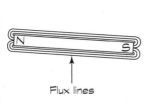

Flux lines

Producing a magnetic field with electricity

As Ørsted discovered, electrical current running through a wire produces a weak magnetic field surrounding the wire. This is why the compass needle moved when the compass was close to Ørsted's circuit. Stop the current from flowing, and the magnetic field disappears. This temporary magnet is electronically controllable — that is, you can turn the magnet on and off by switching current on and off — and it's known as an *electromagnet.*

With the current on, the lines of force encircle the wire and are spaced evenly along the length of the wire, as shown in Figure 5-2. Picture a roll of paper towels with a wire running through its exact center. If you pass current through the wire, invisible flux lines will wrap around the wire along the surface of the roll, and along similar "rings" around the wire at various distances from the wire. The strength of the magnetic force decreases as the flux lines get farther away from the wire. If you wind the current-carrying wire into a uniform coil of wire, the flux lines align and reinforce each other: You've strengthened the magnetic field.

Figure 5-2:
Current
flowing
through a
wire pro-
duces a
weak mag-
netic field
around the
wire.

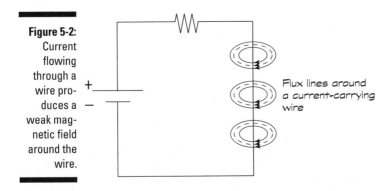

Flux lines around
a current-carrying
wire

Inducing current with a magnet

Hmmm . . . if electricity running through a wire produces a magnetic field, what happens if you place a closed loop of wire near a permanent magnet? Actually, nothing happens — unless you move the magnet. A moving magnetic field will *induce* a voltage across the ends of the wire, causing current to flow through the wire. *Electromagnetic induction* seems to make current magically appear — without any direct contact with the wire. The strength of the current depends on a lot of things, such as the strength of the magnet, the number of flux lines intercepted by the wire, the angle at which the wire cuts across flux lines, and the speed of the magnet's motion. You can increase your chances of inducing a strong current by wrapping the wire into a coil and placing the magnet through the center (*core*) of the coil. The more turns of wire you wrap, the stronger the current will be.

Suppose you place a strong permanent magnet in the center of a coil of wire that is connected as in Figure 5-3. If you move the magnet up, current is induced in the wire and flows in one direction. If you move the magnet down, current is also induced, but it flows in the other direction. By moving the magnet up and down repeatedly, you can produce an alternating current (AC) in the wire. Alternatively, you can move the *wire* up and down around the magnet, and the same thing will happen. As long as there is relative motion between the wire and the magnet, current will be induced in the wire.

Figure 5-3:
Moving a
magnet
inside a
coil of wire
induces a
current in
the wire.

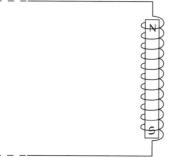

Many power plants generate AC by rotating a conductor inside a strong horseshoe-shaped magnet. The conductor is attached to a rotating turbine, which turns as water or steam applies pressure to its fins. As the conductor makes one full rotation inside the magnet, the magnet pulls electrons first in one direction and then in the other direction, producing alternating current.

Introducing the Inductor: A Coil with a Magnetic Personality

 The circuit symbol for an inductor looks like this. An *inductor* is a passive electronic component made from a coil of wire wrapped around a core — which could be air, iron, or ferrite (a brittle material made from iron). Iron-based core materials increase the strength of the magnetic field induced by current several hundred times. Inductors are sometimes known as *coils, chokes, electromagnets,* and *solenoids,* depending on how they're used in circuits.

REMEMBER

If you pass current through an inductor, you create a magnetic field around the wire. If you *change* the current, increasing it or decreasing it, the magnetic flux around the coil changes, and a voltage is induced across the inductor. That voltage, sometimes called *back voltage,* causes a current flow that opposes the main current. This property of inductors is known as *self-inductance,* or simply *inductance.*

Measuring inductance

Inductance, symbolized by L, is measured in units called *henrys* (named for Joseph Henry, a New Yorker who liked to play with magnets and discovered the property of self-inductance). An inductance of one henry (abbreviated H) will induce one volt when the current changes its rate of flow by one ampere per second. Naturally, one henry is much too large for everyday electronics, so you're more likely to hear about millihenrys (mH) — not because Joe's wife's name was Milly, but because inductance measured in *thousandths* of a henry is more commonplace. You'll also run across microhenrys (μH), which are millionths of a henry.

Opposing current changes

In Figure 5-4, a DC voltage is applied to a resistor in series with an inductor. If there were no inductor in the circuit, then a current equal to V_{supply}/R would flow instantaneously as soon as the DC voltage is switched on. However, introducing an inductor affects what happens to the current flowing in the circuit.

When the DC voltage is first switched on, the current that starts to flow induces a magnetic field around the coils of the inductor. As the current increases (which it's trying to do instantaneously), the strength of the magnetic field increases proportionally. Because the magnetic field is changing, it induces a back voltage which in turn induces a current in the coiled wire *in the opposite direction* to that of the current already flowing from the voltage source. The inductor seems to be trying to prevent the source current from changing too quickly, and the effect is that the current doesn't increase instantaneously. This is why inductors are said to "oppose changes in current."

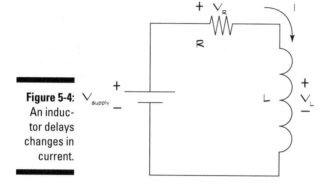

Figure 5-4:
An induc-
tor delays
changes in
current.

The current induced in the coil reduces the strength of the expanding magnetic field a bit. As the source current keeps rising, the magnetic field continues to expand (but more and more slowly), and current opposing the source current continues to be induced (but it gets smaller and smaller). The cycle continues, until finally the overall current settles down to a steady DC. When current reaches a steady level, the magnetic field no longer changes — and the inductor ceases to affect the current in the circuit.

The overall effect is that it takes a finite amount of time for the current flowing through the inductor to reach a steady DC value. (The specific amount of time it takes depends on a few things, such as the characteristics of the inductor and the size of the resistor in the circuit. See "Calculating the RL time constant" later in this chapter.) When this happens, the current flows freely through the inductor, which acts like a simple wire (commonly referred to as a *short circuit,* or simply a *short*) so $V_L = 0$ V and the steady-state current is determined by the source voltage and the resistor according to Ohm's Law ($I = V_{supply}/R$).

If you then remove the DC voltage source and connect the resistor across the inductor, current will flow for a short period of time, with the inductor again opposing the sudden drop in current, until finally the current settles down to zero and the magnetic field disappears.

From an energy perspective, when you apply a DC source to an inductor, it stores electrical energy in a magnetic field. When you remove the DC source and connect a resistor across the inductor, the energy is transferred to the resistor, where it dissipates as heat. Inductors store electrical energy in magnetic fields. A real inductor — as opposed to a theoretical "ideal inductor" — exhibits a certain amount of resistance and capacitance in addition to inductance, due to the physical properties of its windings and core material as well as the nature of magnetic fields. Consequently, an inductor can't retain electrical energy for very long (as a capacitor can) because energy is lost through heat dissipation.

To help you understand inductors, think about water flowing through a pipe with a turbine in it. When you first apply water pressure, the fins of the turbine obstruct the flow, applying back pressure on the water. As the fins start to turn, they apply less back pressure, so the water flows more easily. If you suddenly remove the water pressure, the fins will keep turning for a while, pulling water along with them, until eventually the fins stop turning and the water stops flowing.

Don't worry about the ins and outs of induced currents, expanding and contracting magnetic fields, and the like. Just remember a few things about inductors:

- ✔ An inductor opposes (resists) changes in current.

- ✔ An inductor acts like an open circuit when DC is first applied (that is, no current flows right away, and the entire source voltage is dropped across the inductor).

- ✔ An inductor eventually acts like a short in DC circuits — that is, when all the magnetic-field magic settles down, the voltage is zero, and the inductor allows the full DC current to pass through.

Keeping up with alternating current (or not!)

When you apply an AC voltage to a circuit containing an inductor, the inductor fights against any changes in the source current. If you keep varying the supply voltage up and down at a very high frequency, the inductor will keep opposing the sudden changes in current. At the extreme high end of the frequency spectrum, no current flows at all because the inductor simply can't react quickly enough to the change in current.

Picture yourself standing between two very tempting dessert platters. You want to go for each of them, but can't decide which to try first. You start out running toward one, but quickly change your mind and turn around and starting running toward the other. Then you change your mind again, so you turn and start racing toward the first one, and so forth. The faster you change your mind, the more you stay put in the middle — not getting anywhere (or any dessert). Those tempting desserts make you act like the electrons in an inductor when a high-frequency signal is applied to the circuit: Neither you nor the electrons make any progress.

Understanding inductive reactance

An inductor's opposition to changing current is known as *inductive reactance*. The faster the current tries to change, the more the inductor resists the change.

In the torn-between-two-desserts example just served up here, if you don't change your mind quite so fast, you can run to the first platter, eat a cannoli, then run to the second platter, eat a cream puff, and so forth. If you change your mind a bit faster (but still not every split second), you may find yourself running toward each platter, getting halfway there, and then turning around and running toward the other platter, getting halfway there, and so on. How far you get depends on how quickly you change your mind. It's the same thing for current flow through an inductor: How far a current gets depends on how quickly the current is changing.

Inductive reactance, symbolized by X_L, is measured in — did you guess? — ohms. Inductive reactance, like capacitive reactance (discussed in Chapter 4), varies with frequency, and so is very different from the fixed resistance of your average resistor. You use this formula for X_L:

$$X_L = 2 \times \pi \times f \times L$$

In this formula, f represents the frequency in hertz (Hz) of the applied AC voltage, L is the inductance in henrys, and π is a constant that can be approximated as 3.14. You can approximate $2 \times \pi$ as 6.28 and simplify the formula as follows:

$$X_L \approx 6.28 \times f \times L$$

You calculate the inductive reactance of a 0.1-mH inductor for a source with a frequency of 1 Hz as follows:

$$X_L \approx 6.28 \times 1 \times 0.0001$$
$$\approx 0.000628 \ \Omega$$

If you raise the frequency to 2 MHz (that's 2,000,000 Hz), the inductive reactance is:

$$X_L \approx 6.28 \times 2,000,000 \times 0.0001$$
$$\approx 1.3 k\Omega$$

Notice that for a low-frequency signal (1 Hz), the inductive reactance was nearly zero, so the inductor looks almost like a short circuit, presenting no opposition to current. For a high-frequency signal (2 MHz), the inductor puts up significant opposition to current flow (1.3kΩ of reactance).

Using Ohm's Law for inductive reactance

You can use Ohm's Law for inductive reactance, as for capacitive reactance, as long as you're careful and apply it for a single frequency at a time. Suppose you apply a 2 MHz AC voltage with a peak value of 5 V to your 0.1-mH inductor. You can calculate the peak current flowing through the inductor by applying Ohm's Law:

$$I_{peak} = \frac{V_{peak}}{X_L}$$

The inductive reactance for a 0.1-mH inductor at 2 MHz is 1.3kΩ (as calculated above), so the peak current is:

$$I_{peak} \approx \frac{5\ V}{1,300\ \Omega}$$
$$\approx 0.0038\ A\ or\ 3.8\ mA$$

Behaving differently depending on frequency (again!)

Like capacitors, inductors in an AC circuit act differently depending on the frequency of the voltage applied to them. Because the current passing through an inductor is affected by frequency, so are the voltage drops across the inductor and other components in the circuit. This frequency-dependent behavior forms the basis for useful functions, such as low-, high-, and band-pass filters.

 Inductors are sort of the alter egos of capacitors. Capacitors oppose voltage changes; inductors oppose current changes. Capacitors block DC and pass more and more AC as frequency increases; inductors pass DC and block more and more AC as frequency increases.

Using Inductors in Circuits

Inductors are used primarily in tuned circuits, to select or reject signals of specific frequencies, and to block (or choke) high-frequency signals, such as eliminating radio frequency (RF) interference in cable transmissions. In audio applications, inductors are also commonly used to remove the 60 Hz hum known as *noise* (often created by nearby power lines).

You probably run into (or over) inductors more often than you think. Many traffic-light sensors — those devices that seem to know there's a car waiting for the light to change — use an inductor to trigger the light to change.

Embedded in the street several yards before the intersection is an inductive loop consisting of several turns of a gigantic coil, roughly six feet in diameter. This loop is connected to a circuit that controls the traffic signal. As you pass over the loop, the steel underbody of your car changes the magnetic flux of the loop. The circuit detects this change — and gives you the green light.

Insulating and shielding inductors

The wire that makes up an inductor must be insulated to prevent unintended short circuits between the turns. Inductors used in most electronic circuits are also *shielded,* or encased in a nonferrous metal can (typically brass or aluminum), to prevent the magnetic lines of flux from infiltrating the neighborhood of other components in a circuit. You use a shielded inductor when you don't want to induce voltages or currents in other circuit elements. You use an unshielded inductor (or coil) when you do want to affect other circuit elements. We discuss the use of unshielded coils in circuits in the section "Influencing the Coil Next Door: Transformers."

Reading inductance values

The value of an inductor is typically marked on its package using the same color-coding technique used for resistors, which you can read about in Chapter 3. You can often find the value of larger inductors printed directly on the components. Smaller-value inductors look a lot like low-wattage resistors; such inductors and resistors even have similar color-coding marks. Larger-value inductors come in a variety of sizes and shapes that depend how they're designed to be used.

Inductors can be either fixed or variable. With either type, a length of wire is wound around a core. The number of turns of the wire, the core material, the wire's diameter, and length of the coil all determine the numerical value of the inductor. Fixed inductors have a constant value; variable inductors have adjustable values. The core of an inductor can be made of air, iron ferrite, or any number of other materials (including your car). Air and ferrite are the most common core materials.

Combining shielded inductors

Chances are, you won't use inductors in the basic electronic circuits you set up, but you may run across circuit diagrams for power supplies and other devices that include multiple inductors. Just in case you do, you should know how to calculate the equivalent inductance of combinations of shielded inductors so you can get a clear picture of how the circuit operates.

Inductors in series add up, just as resistors do:

$$L_{series} = L1 + L2 + L3 \ldots$$

Like resistors, inductors in parallel combine by adding the reciprocals of each individual inductance, and then taking the reciprocal of that sum. (You may remember from math class that the reciprocal is the multiplicative inverse of a number, or the number that you multiply by so that the result equals 1. So for any integer, x, $1/x$ is its reciprocal.)

$$L_{parallel} = \cfrac{1}{\dfrac{1}{L1} + \dfrac{1}{L2} + \dfrac{1}{L3} \ldots}$$

Another way to express the equation above is:

$$\cfrac{1}{L_{parallel}} = \dfrac{1}{L1} + \dfrac{1}{L2} + \dfrac{1}{L3} \ldots$$

If you have just two inductors in parallel, you can simplify this as follows:

$$L_{parallel} = \dfrac{L1 \times L2}{L1 + L2}$$

Filtering signals with inductors

Remember: Inductors are like the alter egos of capacitors. It turns out that you can use inductors in filter circuits to do exactly the opposite of what capacitors do in filter circuits, as we discuss in Chapter 4.

Figure 5-5 shows an *RL circuit,* which is simply a circuit that contains both a resistor and an inductor, with the output voltage, V_{out}, defined as the voltage dropped across the inductor.

The lower the frequency of the input voltage, the more the inductor is able to react to changes in current, so the more the inductor looks like a short, passing current with little or no opposition. As a result, for low-input frequencies, the output voltage is nearly zero. The higher the frequency of the input voltage, the more the inductor fights the change in current and the less successful the input voltage is in pushing current through the circuit. As a result, for high-input frequencies, very little voltage is dropped across the resistor (because $V_R = I \times R$, and I is very low), so most of the input voltage is dropped across the inductor, and the output voltage is nearly the same as the input voltage.

This RL circuit is configured as a *high-pass filter,* because it "allows" high-frequency input signals to "pass" through to the output, while blocking DC and low frequencies from getting through.

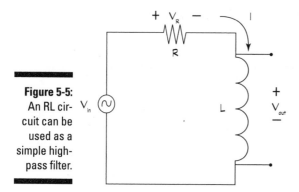

Figure 5-5:
An RL cir-
cuit can be
used as a
simple high-
pass filter.

If you reverse the roles of the resistor and the inductor in Figure 5-5, and define the output voltage to be the voltage dropped across the resistor, you have a *low-pass filter*. For lower frequencies, because the inductor behaves more and more like a short, little or no voltage is dropped across it, so nearly all of the input voltage is dropped across the resistor; at higher frequencies, the inductor acts more and more like an open circuit, allowing less and less current to flow through the circuit, and thus, through the resistor — so the output voltage is *attenuated* (decreased) dramatically.

Calculating the RL time constant

You can calculate the amount of time (in seconds) it takes for an inductor's induced voltage to reach roughly two-thirds of its value by using the RL time constant, T. The formula looks like this:

$$T = {}^L\!/_R$$

Just as the RC time constant in RC circuits (which you can read about in Chapter 4) gives you an idea of how long it takes a capacitor to charge up to its full capacity, so the RL time constant helps you figure out how long it will take for an inductor to fully conduct a DC current: Direct current settles down to a steady value after roughly five RL time constants. You can also use the RL time constant to calculate the *filter cutoff frequency* (the frequency at which a filter starts to affect an input signal) as follows:

$$F_c = (2 \times \pi \times T)^{-1}$$
$$= (2 \times \pi \times L)^{-1}$$

Now Introducing Impedance!

As mentioned earlier in this chapter, an inductor's opposition to changing current is known as inductive reactance — just as a capacitor's opposition to changing current is known as capacitive reactance (more about that in Chapter 4). Although reactance and resistance both involve opposing alternating current (AC), they are not the same; resistance is constant over all frequencies, whereas reactance varies with frequency. In circuits that have reactance *and* resistance, for instance, the RL circuit in Figure 5-5, you may need to know the total opposition to alternating current in the circuit for an input signal at a certain frequency.

Both resistance and reactance are measured in ohms, so you may think that you can simply add the inductive reactance, X_L, to the resistance, R, to get the total opposition to current (like adding up series resistances) — but you can't. The reason you can't has to do with the time it takes for an inductor (and similarly, a capacitor) to *react to* changes in the circuit. The good news is that there is a way to calculate the total opposition to current taking place in a circuit at a particular frequency.

Impedance is the total opposition a circuit provides to varying current for a given frequency. (This is analogous to the equivalent resistance of a purely resistive circuit, as discussed in Chapter 3, which takes into account all the individual resistances in the circuit.) You use the symbol Z to represent impedance. Impedance takes into account the total resistance *and* total reactance of a circuit.

You can find total impedance for a series RL circuit by using this formula:

$$Z_L = \sqrt{R^2 + X_L{}^2}$$

Similarly, for a series RC circuit, the total impedance is

$$Z_C = \sqrt{R^2 + X_C{}^2}$$

So how do you *use* those scary-looking formulas? Well, if you want to use Ohm's Law to calculate the current through a series RL or RC circuit when you apply an input signal of a particular frequency, you can calculate the total impedance of the circuit at that frequency, and then apply Ohm's Law, using the impedance to calculate the peak value of the current passing through the circuit for that one particular frequency:

$$I_{peak} = \frac{V_{peak}}{Z_L}$$

If you have a circuit with a resistor, capacitor, and inductor all in series (one type of *RLC circuit*), the formula for impedance is more complicated, because the total reactance in the circuit (X_T) is the *difference* between the inductive reactance, X_L, and the capacitive reactance, X_C ($X_T = X_L - X_C$). For parallel arrangements of resistors and inductors, or resistors and capacitors, the formula for impedance is even more complicated, but relax; we won't take you down that path in this book.

Tuning in to Radio Broadcasts

Inductors are natural low-pass filters, and capacitors are natural high-pass filters, so what happens when you put the two in the same circuit? As you might guess, inductors and capacitors are often used together in tuning circuits, to "tune in" a specific radio station's broadcast frequency.

Resonating with RLC circuits

Look at the RLC circuit in Figure 5-6. The total impedance of this circuit can be found using the following *really* scary formula:

$$Z = \sqrt{R^2 + X^2}$$
$$= \sqrt{R^2 + \left(X_L - X_C\right)^2}$$

Don't worry about how scary that formula looks. Just notice one thing about that formula: If $X_L = X_C$, then the total impedance becomes the square root of R^2, which is just R, or just the resistance. In other words, when $X_L = X_C$, the inductive reactance and the capacitive reactance *cancel each other out;* it's almost as if there is just a resistor in the circuit. This happens at exactly one frequency, known as the *resonant frequency.* The resonant frequency is the value of frequency, f, that makes $X_L = X_C$ for a given combination of inductance (L) and capacitance (C).

So what is the significance of the resonant frequency? Well, something special happens at that frequency that doesn't happen at any other frequency. At frequencies above and below the resonant frequency, there is some overall reactance in the circuit adding to the resistor's opposition to current flow. For very low frequencies, the capacitor exhibits a high reactance; for very high frequencies, the inductor exhibits a high reactance. So for low-frequency signals, the capacitor squelches current flow; for high frequencies, the inductor squelches current flow. At the resonant frequency, the overall reactance is zero, and it looks like only the resistance opposes current flow. The circuit is said to *resonate* at that particular frequency, and so is known as a *resonant circuit*. Figure 5-7 shows a frequency plot of the current passing through the circuit; notice that the current is highest at the resonant frequency.

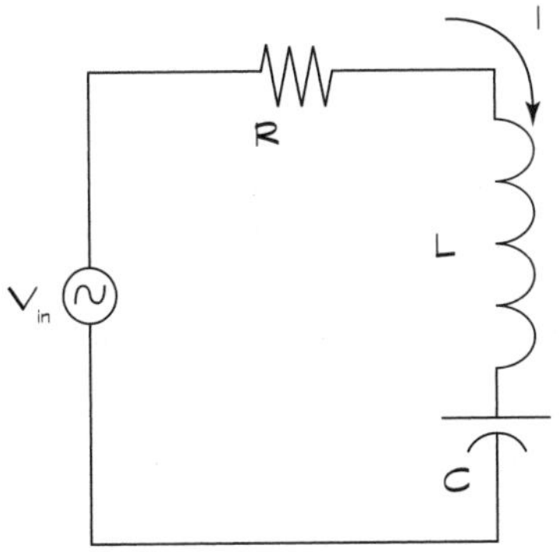

Figure 5-6:
An RLC circuit has a resonant frequency, at which the total reactance is zero.

Radio receivers use RLC circuits to allow just one frequency to pass through the circuit. This is known as "tuning in" to the frequency and used this way, the circuit is known as a *tuning circuit*. A variable capacitor is used to adjust the resonant frequency, so you can tune in different stations broadcasting at different frequencies. The knob that allows the capacitance to be changed is attached to the tuning-control knob on your radio.

By shifting components around a bit, placing the inductor in parallel with the capacitor, you create a circuit that produces the *minimum* current at the resonant frequency. This sort of resonant circuit "tunes out" that frequency, allowing all others to pass, and is used to create *band-stop filters*. You might find such a circuit filtering out the 60 Hz hum that electronic equipment sometimes picks up from a nearby power line.

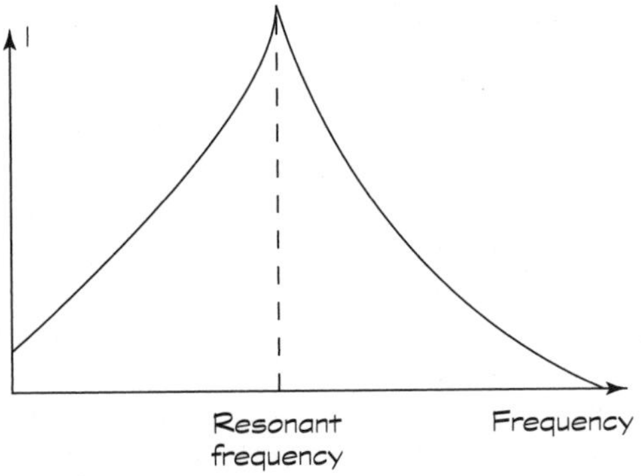

Figure 5-7:
The current in a series RLC circuit is highest at the resonant frequency.

Ensuring rock-solid resonance with crystals

This is the circuit symbol for a crystal, abbreviated XTAL. If you slice a quartz crystal in just the right way, mount two leads to it, and enclose it in a hermetically sealed package, you've created a single component that acts like an RLC combo in an RLC circuit, resonating at a particular frequency. *Quartz crystals*, or simply *crystals*, are used in circuits to generate an electrical signal at a very precise frequency.

Crystals work because of something called the *piezoelectric effect:* If you apply a voltage in just the right way across a quartz crystal, it vibrates at a specific frequency, known as the resonant frequency. If you then remove the applied voltage, the crystal continues to vibrate until it settles back to its previous shape. As it vibrates, it generates a voltage at the resonant frequency.

You may be familiar with piezoelectric guitar pickups, which use crystals to convert the mechanical vibrations generated by guitar strings into electrical signals, which are then amplified. And if you pre-date compact disc (CD) technology, you may be interested to know that phonograph needles relied on the piezoelectric effect to convert the ups and downs of a vinyl record track into electrical energy.

The frequency at which a crystal resonates depends upon its thickness and size, and you can find crystals with resonant frequencies ranging from a few tens of kilohertz to tens of megahertz. Crystals are more precise and more reliable than combinations of capacitors and inductors, but there's a catch: They're usually more expensive. You will find crystals used in circuits called *oscillators* to generate electric signals at a very precise frequency. Oscillators are responsible for the ticks and tocks that control quartz wristwatches and digital integrated circuits (which we discuss in Chapter 7), and for controlling the accuracy of radio equipment.

A quartz crystal is accurate to within roughly 0.001 percent of its stated resonant frequency. (That why they're worth paying some extra bucks for!) You may also hear of ceramic resonators, which work the same way but cost less and are not as accurate as quartz. Ceramic resonators have a 0.5 percent frequency tolerance — meaning that the actual resonant frequency can vary by as much as 0.5 percent above or below its stated resonant frequency — and are used in many consumer-electronics devices, such as TVs, cameras, and toys.

Influencing the Coil Next Door: Transformers

Inductors used in tuning circuits are shielded so that the magnetic field they produce doesn't interact with other circuit components. Unshielded coils are sometimes placed close to one another for the express purpose of allowing their magnetic fields to interact. In this section, we describe how unshielded coils interact — and how you can exploit their interaction to do some useful things with an electronic device known as a *transformer*.

Letting unshielded coils interact

When you place two unshielded coils near each other, the varying magnetic field created as a result of passing AC through one coil induces a voltage in that coil *as well as in the other coil*. *Mutual inductance* is the term used to describe the effect of inducing a voltage in another coil, while *self-inductance* refers to the effect of inducing a voltage in the same coil that produced the varying magnetic field in the first place. The closer the coils, the stronger the interaction. Mutual inductance can add to or oppose the self-inductance of each coil, depending on how you match up the north and south poles of the inductors.

If you have an unshielded coil in one circuit, and place it close to an unshielded coil in another circuit, the coils will interact. By passing a current through one coil, you will cause a voltage to be induced in the neighboring coil — even though it is in a separate, unconnected circuit. This is known as *transformer action*.

A *transformer* is an electronic device that consists of two coils wound around the same core material in such a way that the mutual inductance is maximized. Current passing through one coil, known as the *primary*, induces a voltage in the other coil, known as the *secondary*. The job of a transformer is to transfer electrical energy from one circuit to another.

The circuit symbols for an air-core transformer and solid-core transformer, respectively, are shown in Figure 5-8.

Figure 5-8:
Circuit
symbols for
an air-core
transformer
and a
solid-core
transformer.

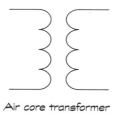

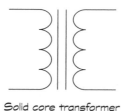

Air core transformer   Solid core transformer

Isolating circuits from a power source

If the number of turns of wire in the primary winding of a transformer is the same as the number of turns in the secondary winding, theoretically, all of the voltage across the primary will be induced across the secondary. This is known as a *1:1 transformer,* because there is a 1:1 (read "one-to-one") relationship between the two coils. (In reality, no transformer is perfect, or *lossless,* and some of the electrical energy gets lost in the translation.)

1:1 transformers are also known as *isolation transformers,* and are commonly used to electrically separate two circuits while allowing power or an AC signal from one to feed into the other. The first circuit typically contains the power source, and the second circuit contains the load. (In Chapter 1, we define the load as the destination for the electrical energy, or the thing you ultimately want to perform work on, such as a speaker diaphragm.) You may want to isolate circuits to reduce the risk of electrical shocks or to prevent one circuit from interfering with the other.

Stepping up, stepping down voltages

If the number of turns in the primary winding of a transformer is not the same as the number of turns in the secondary winding, the voltage induced in the secondary will be different from the voltage across the primary. The two voltages will be proportional to each other, with the proportion determined by the ratio of the number of turns in the secondary to the number of turns in the primary, as follows:

$$\frac{V_s}{V_p} = \frac{N_s}{N_p}$$

In this equation, V_S is the voltage induced in the secondary, V_P is the voltage across the primary, N_S is the number of turns in the secondary, and N_P is the number of turns in the primary.

Say, for instance, that the secondary consists of 200 turns of wire — twice as many as the primary, which consists of 100 turns of wire. If you apply an AC voltage with a peak value of 50 V to the primary, the peak voltage induced across the secondary will be 100 V, or twice the value of the peak voltage across the primary. This type of transformer is known as a *step-up trans-former,* because it steps up the voltage from the primary to the secondary.

If, instead, the secondary consists of 50 turns of wire, and the primary consists of 100 turns, the same AC signal applied to the same primary has a different result: The peak voltage across the secondary would be 25 V, or half the primary's voltage. This is known as a *step-down transformer,* for obvious reasons.

In each case, the power applied to the primary winding is transferred to the secondary. Because power is the product of voltage and current ($P = V \times I$), the *current* induced in the secondary winding is inversely proportional to the *voltage* induced in the secondary. So a step-up transformer steps up the voltage while stepping down the current; a step-down transformer steps down the voltage while stepping up the current.

Step-up and step-down transformers are used in electrical power-transmission systems. Electricity generated at a power plant is stepped up to voltages of 110 kV (1 kV = 1,000 V) or more, transported over long distances to a substation, and then stepped down to lower voltages for distribution to customers.

Chapter 6

The Wide World of Semiconductors

Semiconductors are at the heart of nearly every major electronic system that exists today, from the programmable pacemaker to the space shuttle. It's amazing to think that teeny-tiny semiconductors are responsible for triggering enormous advances in modern medicine, space exploration, industrial automation, home entertainment systems, communications, and a slew of other industries.

Semiconductor diodes and transistors can be made to conduct or block electric current — depending on how you control them electrically. They make it possible to allow current to flow in one direction, but not the other, and to amplify tiny signals — tasks that your average passive electronic component cannot perform.

This chapter tours the insides of semiconductor materials, shows how to make them conduct current, and explores ways to combine semiconductors to create diodes and transistors. Next you get a look at the valve-like behavior of diodes (and get a handle on how to exploit that behavior in circuits), followed by a peek at how transistors work and why they're so popular. Then, *voilà* — the stage is set for modern-day electronics.

Are We Conducting or Aren't We?

Somewhere between insulators and conductors are materials that can't seem to make up their minds about whether to hold on to their electrons or let them roam freely. These *semiconductors* behave like conductors under some conditions and insulators under other conditions, giving them unique capabilities. With a device made from semiconductor materials, such as silicon and germanium, you can precisely control the flow of electrical charge carriers in one area of the device by adjusting a voltage in another area of the device.

You use semiconductor diodes to allow current to flow in one direction while blocking its flow in the other direction, just like a check valve. Transistors enable you to switch electron flow (current) on and off, and to control the strength of a larger current by adjusting a smaller current. These current-control capabilities enable lots of complex electronic functions, including amplification of electrical signals, *rectification* (converting AC to DC), and current steering. And the microscopic size and low power consumption qualities of semiconductor devices enable you to converge all those complex control functions into the tiny low-power devices that have revolutionized modern-day electronics.

The atoms of semiconductor materials align themselves in a structured way, forming a very regular, three-dimensional pattern — a crystal — as shown in Figure 6-1. Atoms within the crystal are held together by a special bond, called a *covalent bond,* with each atom sharing its outermost electrons (known as *valence electrons*) with its neighbors. (Bonding — it's a wonderful thing!) This is sort of like next-door neighbors sharing a common driveway: Each one behaves as if the driveway belongs to him (except after a major snowstorm).

It's precisely because of this unique touchy-feely bonding and sharing of electrons that the semiconductor crystal acts like an insulator most of the time. Each atom thinks it has more valence electrons than it really has — and those electrons behave as part of a large, happy family of electrons, with no need to go wandering off to another atom. (This is very different from a typical conductor atom, which often has just one valence electron, and that lone electron is very susceptible to leaving home and drifting around.) But there is something you can do to a pure semiconductor that will change the electrical properties of the material: Dope it. (No, don't muddle its brain with illegal substances; just add a material that makes the electrons flow differently. More about that in a minute.)

The rest of this section tells you a little bit about the underlying theory of semiconductor physics. This information is useful, but you don't really need it to understand how semiconductor components work. If you just skip ahead to the section, "Forming a Junction Diode," later in this chapter, you'll still read what you need in order to use semiconductors in circuits.

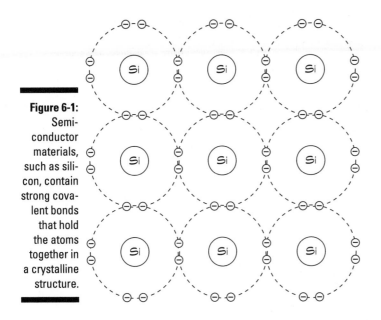

Figure 6-1:
Semi-
conductor
materials,
such as sili-
con, contain
strong cova-
lent bonds
that hold
the atoms
together in
a crystalline
structure.

Doping semiconductors

If you add impurities (meaning different types of atoms, never mind the dirt and dander) to a pure semiconductor material such as silicon, you upset the bonding applecart. This process is known as *doping,* and the impurities are called *dopants.* Arsenic and boron are two dopants commonly used to dope pure silicon.

Dopants are no dopes; they try to masquerade as one of the crystal's atoms, attempting to bond with the other atoms, but they are just different enough to stir things up a bit. For instance, an atom of arsenic has one more outer electron than an atom of silicon. When you add a small amount of arsenic to a bunch of silicon atoms, each arsenic atom muscles its way in, bonding with the silicon atoms, but leaving its "extra" electron drifting around through the crystal. Even though the doped material is electrically neutral, it now contains a bunch of "free" electrons wandering aimlessly — making it much more conductive. By doping the silicon, you change its electrical properties: Wherever the dopant is added, the silicon becomes more conductive.

Another way to dope semiconductors is to use materials like boron, in which each atom has one *fewer* valence electron than does a silicon atom. For every boron atom you add to a silicon crystal, you get what is known as a *hole* in the crystalline structure where an outer electron should be. Wherever there is a hole in the structure, the bond holding the atoms together is so strong, it will "steal" an electron from another atom to fill the hole, leaving a hole some-where else, which then gets filled by another electron, and so forth. You can think of this as the hole moving around inside the crystal, as illustrated in

Figure 6-2. (Really, it's electrons that are moving, but it looks like the position of the hole keeps moving.) Because each hole represents a missing electron, the movement of holes has the same effect as a flow of positive charges.

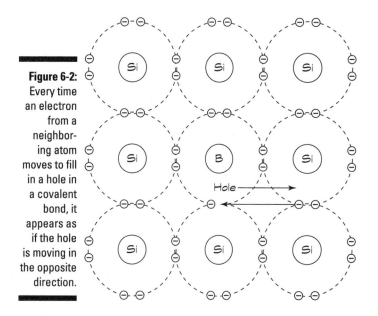

Figure 6-2:
Every time an electron from a neighboring atom moves to fill in a hole in a covalent bond, it appears as if the hole is moving in the opposite direction.

Impurities that free up electrons (negative charges) to move through a semiconductor are called *donor dopants,* and the doped semiconductor is known as an *N-type semiconductor.* Arsenic is a typical donor dopant.

Impurities (such as boron) that free up holes (like positive charges) to move through a semiconductor are called *acceptor dopants,* and the doped semiconductor is known as a *P-type semiconductor.* Boron is a typical acceptor dopant.

Combining N-types and P-types to create components

If you apply a voltage source across either an N-type or a P-type semiconductor, electrons move through the material and current flows from the negative voltage toward the positive voltage. (For P-type semiconductors, you'll hear this described as a movement of holes from the positive voltage toward the negative voltage.) So far, so good: The doped semiconductors are simply acting like conductors, and you could just as easily use copper wire to achieve the same effect.

Diving into a pn-junction

To really understand the reasons why current flows or does not flow when you apply voltages across a pn-junction, you need to dive a bit deeper into the physics that lies beneath the surface of the junction. Although we won't dive too deep here, we will tell you that it's all about which way the holes are pushed in the P-type material, which way the electrons are pushed in the N-type material, and how holes and electrons sometimes re-combine at the junction.

Even with no external voltage applied, there is still a small voltage difference across the junction. That voltage is caused by the holes and electrons that meet up at the junction, cross over to the other side, and re-combine (meaning that an electron fills a hole). This leaves the area on either side of the junction with a net charge: On the P-side of the junction, the net charge is negative; on the N-side, the net charge is positive. So there's a difference in charge (that is, a small voltage) across the junction. That voltage opposes the further flow of holes and electrons across the junction.

By connecting the positive terminal of a battery to the P-type material and the negative terminal of the battery to the N-type material, you push holes through the P-type material toward the junction, and you push electrons through the

N-type material toward the junction. If you send enough of those two types of charges toward the junction, you will overcome the small voltage that exists there already; holes will cross into the N-type material, and electrons will cross into the P-type material. The external voltage supply keeps pushing charges toward the junction, so charges keep crossing the junction. The net effect of holes moving one way and electrons moving the other way is a current. Conventional current (movement of positive charges) moves from the positive battery terminal through the P-type material, across the junction and through the N-type material toward the negative battery terminal.

If you connect the battery the other way, the negative terminal attracts holes from the P-type material, pulling them away from the junction, and the positive terminal attracts electrons from the N-type material, pulling them away from the junction. This in effect strengthens the already-existing voltage across the junction, making it even harder for current to flow. If the externally applied voltage is strong enough, eventually current can be made to flow — in the reverse direction across the junction. The voltage at which this happens is called the *breakdown voltage,* and this breakdown process is what makes a *Zener diode* work.

Things start to get tricky when you fuse together an N-type and a P-type semiconductor and apply a voltage across this *pn-junction.* Whether or not a current flows depends on which way you apply the voltage. If you connect the positive terminal of a battery to the P-type material, and the negative terminal to the N-type material, current will flow (as long as the applied voltage exceeds a certain minimum). But if you reverse the battery, current will not flow (unless you apply a very large voltage).

Exactly how these N-type and P-type semiconductors are combined determines what sort of semiconductor device they become — and how they allow current to flow (or not) when voltage is applied. The pn-junction is the foundation for *solid-state* electronics, which involves electronic devices made of solid, non-moving materials, rather than vacuum tubes or devices with movable parts (such as the mechanical disks and tapes used for storing computer information). Semiconductors have largely replaced vacuum tubes in electronics.

Forming a Junction Diode

A semiconductor *diode* is a two-terminal electronic device that consists of a single pn-junction. The circuit symbol for a diode is shown here. Diodes act like one-way valves, allowing current to flow in only one direction when a voltage is applied to them. This capability is sometimes referred to as the *rectifying* property.

You refer to the P-side of the pn-junction in a diode as the *anode,* and the N-side as the *cathode.* In the circuit symbol above, the anode is on the left (broad end of the arrowhead) and the cathode is on the right (short vertical line segment). Most diodes allow current to flow from the anode to the cathode. (Zener diodes are an exception; for details, flip ahead in this chapter to "Regulating voltage with Zener diodes.") You can think of the junction within a diode as a hill and the current as a ball you are trying to move from one side of the hill to the other. It's easy to push the ball down the hill (from anode to cathode), but very difficult to push the ball up the hill (from cathode to anode).

Diodes are cylindrical, like resistors, but aren't quite as colorful as resistors. Most diodes sport a stripe or other mark at one end, signifying the cathode. Figure 6-3 shows some diodes.

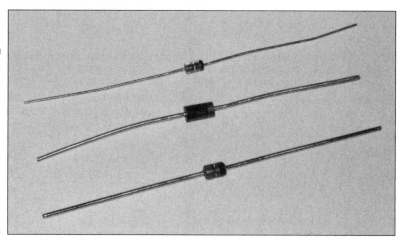

Figure 6-3: Diodes are similar in size and shape to resistors, but include just one stripe indicating the cathode.

Biasing the diode

In electronics, the term *bias* refers to a steady DC voltage or current applied to an electronic device or circuit to get it to operate a certain way. Devices like diodes and transistors (which we cover later in this chapter) are nonlinear devices: That is, the relationship between voltage and current in these devices is not constant. It varies across different ranges of voltages and current. They're not like resistors, which exhibit a linear (constant) relationship between voltage and current.

When you bias a diode, you apply a voltage, known as the *bias voltage,* across the diode (from anode to cathode) so that the diode either allows current to flow from anode to cathode or blocks current from flowing. These two basic operating modes of a standard diode are known as *forward-bias* (conducting) and *reverse-bias* (non-conducting).

You *forward-bias* a diode — meaning, get it to conduct current — by applying a high enough positive voltage from anode to cathode so that the diode "turns on" (conducts current). This minimum turn-on voltage is known as the *forward voltage,* and its value depends on the type of diode. A typical silicon diode has a forward voltage of about 0.6 V to 0.7 V, whereas forward voltages for light-emitting diodes (LEDs) range from about 1.5 V to 4.6 V (depending on the color). (Check the ratings on the particular diodes you use in circuits.) When the diode is forward-biased, current, known as *forward current,* flows easily across the pn-junction, from anode to cathode. You can increase the amount of current flowing through the diode (up to the maximum current it can safely handle), but the forward voltage drop won't vary that much.

Figure 6-4 shows a forward-biased diode allowing current to flow through a lamp.

Figure 6-4:
The battery forward-biases the diode in this circuit, allowing current to flow through a lamp.

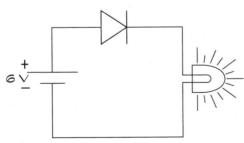

You *reverse-bias* a diode when you apply a *reverse voltage* (a negative voltage from anode to cathode) across the diode, prohibiting current from flowing, as shown in Figure 6-5. (Actually, a small amount of current, in the μA range, will flow.) If the reverse-biased voltage exceeds a certain level (usually 50 V or more), the diode "breaks down" and *reverse current* starts flowing from cathode to anode. The reverse voltage at which the diode breaks down is known as the *peak reverse voltage* (PRV), or *peak inverse voltage* (PIV).

You usually don't purposely reverse-bias a diode (except if you're using a Zener diode, which we describe in the section, "Regulating voltage with Zener diodes," later in this chapter). You may accidentally reverse-bias a diode by orienting it incorrectly in a circuit (see the section "Which end is up?" later in this chapter), but don't worry: You won't harm the diode and you can simply re-orient it. (But if you exceed the PRV, you may allow too much reverse current to flow, which can damage other circuit components.)

If no voltage or a low voltage (less than the forward voltage) is applied across a diode, it is *unbiased.* (That doesn't mean the diode lacks prejudice; it just means that you haven't taken action on the diode yet.)

Figure 6-5:
Because the diode in this circuit is reverse-biased, it acts like a closed valve, prohibiting the flow of current through the circuit.

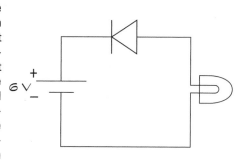

Conducting current through a diode

After current starts flowing through a diode, the forward voltage drop across the diode remains fairly constant — even if you increase the forward current. For instance, most silicon diodes have a forward voltage of between 0.6 V and 0.7 V over a wide range of forward currents. If you're analyzing a circuit that contains a silicon diode (such as the circuit in Figure 6-4), you can assume that the voltage drop across the diode is about 0.7 V — even if you increase the source voltage from 6 V to 9 V. Increasing the source voltage increases the current through the circuit, but the diode voltage drop remains the same, so the increased source voltage is dropped across the lamp.

Of course, every electronic component has its limits. If you increase the current through a diode too much, you'll generate a lot of heat within the diode. At some point, the junction will be damaged from all that heat, so you have to be careful not to turn up the source voltage too high.

Rating your diode

Most diodes don't really have values like resistors and capacitors. A diode simply does its thing in controlling the on/off flow of electrons, without altering the shape or size of the electron flow. But that doesn't mean all diodes are the same. Standard diodes are rated according to two main criteria: peak reverse voltage (PRV) and current. These criteria guide you to choosing the right diode for a particular circuit, as follows:

- The **PRV rating** tells you the maximum reverse voltage the diode can handle before breaking down. For example, if the diode is rated at 100 V, you shouldn't use it in a circuit that applies more than 100 V to the diode. (Circuit designers build in considerable "headroom" above the PRV rating to accommodate voltage spikes and other conditions. For instance, it's common practice to use a 1,000 V PRV rectifier diode in power supply circuits running on 120 VAC.)

- The **current rating** tells you the maximum forward current the diode can withstand without sustaining damage. A diode rated for 3 A can't safely conduct more than 3 A without overheating and failing.

Identifying with diodes

Most diodes originating in North America are identified by five- or six-digit codes that are part of an industry-standard identification system. The first two digits are always "1N" for diodes; the "1" specifies the number of pn-junctions, the "N" signifies semiconductor, and the remaining three or four digits indicate specific features of the diode. A classic example is the series of rectifier diodes identified as 1N40*xx,* where *xx* could be 00, 01, and so forth through 08. They are rated at 1 amp with PRV ratings ranging from 50 to 1,000 V, depending on the *xx* number. For instance, the 1N4001 rectifier diode is rated at 1 A and 50 V, and the 1N4008 is rated at 1 A and 1,000 V. Diodes in the 1N54*xx* series have a 3-A rating with PRV ratings from 50 to 1,000 V. You can readily find such information in any catalog of electronic components or cross-reference book of diode data, generally found online. (A *cross-reference book* tells you what parts can be substituted for other parts, in case a part specified in a circuit diagram is not available from your chosen source.)

Just to make things interesting (not to mention confusing), some diodes use the same color-coding scheme on their packaging as resistors, but instead of translating the code into a value (such as resistance), the color code simply gives you the semiconductor identification number for the diode. For instance, the color sequence "brown-orange-red" indicates the numerical sequence "1-3-2" so the diode is a 1N132 germanium diode. (Refer to the resistor color code chart in Chapter 3.)

Which end is up?

When you use a diode in a circuit, it's extremely important to orient the diode the right way (more about that in a minute). The stripe or other mark on the diode package corresponds to the line segment in the circuit symbol for a diode: Both indicate the cathode, or negative terminal, of the diode.

You can also determine which end is what by measuring the resistance of the diode (before you insert it in your circuit) with an ohmmeter or multimeter (which we discuss in Chapter 12). The diode has a low resistance when it's forward-biased, and a high resistance when it's reverse-biased. By applying the positive lead of your meter to the anode and the negative lead to the cathode, your meter is essentially forward-biasing the diode (because when used to measure resistance, a multimeter applies a small voltage across its leads). You can measure the resistance twice, applying the leads first one way and then the other way. The lower measurement result indicates the forward-biased condition.

Diodes are like one-way valves, allowing current to flow in one direction only. If you insert a diode backward in a circuit, either your circuit won't work at all (because no current will flow) or you may damage some components (because you may exceed the peak reverse voltage [PRV] and allow current to flow in reverse — which can damage components such as electrolytic capacitors). Always note the orientation of the diode when you use it in a circuit, double-checking to make sure you have it right!

Using Diodes in Circuits

You'll find several different flavors of semiconductor diodes designed for various applications in electronic circuits.

Rectifying AC

Figure 6-6 shows a circuit with a silicon diode, a resistor, and an AC power source. Notice the orientation of the diode in the circuit: Its anode (positive end) is connected to the power source. The diode conducts current when it's forward-biased, but not if it's reverse-biased. When the AC source is positive (and provides at least 0.7 V to forward-bias the silicon diode), the diode conducts current; when the AC source is less than 0.7 V, the diode does not conduct current. The output voltage is a clipped version of the input voltage; only the portion of the input signal that is greater than 0.7 V passes through to output.

Figure 6-6:
The diode in this circuit "clips off" the negative half of the AC source voltage.

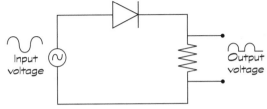

If the diode orientation is reversed in the circuit, the opposite happens: Only the negative part of the input voltage is passed through to the output:

- ✔ When the input voltage is positive, the diode is reverse-biased and no current flows.

- ✔ When the input is sufficiently negative (at least –0.7 V), the diode is forward-biased and current flows.

Diodes used this way — to convert AC current into varying DC current (it's DC because the current is flowing in one direction only, but it isn't a constant current) — are called *rectifier diodes,* or just *rectifiers.* They usually handle currents ranging from several hundred milliamps to a few amps — much higher strengths than general-purpose *signal diodes* are designed to handle (those currents only go up to about 100 mA). You'll see rectifiers used in two major ways:

- ✔ **Half-wave rectification:** Using a single rectifier diode to clip an AC signal is known as *half-wave rectification* because it converts half the AC signal into DC.

✔ **Full-wave rectification:** By arranging four diodes in a circuit known as a *bridge rectifier,* you can convert both the "ups" and the "downs" of an AC voltage into just "ups" (as shown in Figure 6-7). This process, known as *full-wave rectification,* is the first stage of circuitry in a *linear power supply,* which converts AC power into a steady DC power supply.

Bridge rectifiers are so popular, you can purchase them as a single four-terminal part, with two leads for the AC input and two leads for the DC output.

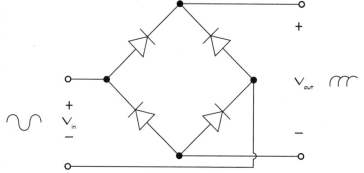

Figure 6-7: In a bridge rectifier, four diodes transform an AC current into a varying DC current.

Regulating voltage with Zener diodes

Zener diodes are special diodes that are meant to break down. They are really just heavily-doped diodes that break down at much lower voltages than standard diodes. When you reverse-bias a Zener diode, and the voltage across it reaches or exceeds its breakdown voltage, the Zener diode suddenly starts conducting current backward through the diode (from cathode to anode). As you continue to increase the reverse-biased voltage beyond the breakdown point, the Zener continues to conduct more and more current — while maintaining a steady voltage drop.

Keep in mind these two important ratings for Zener diodes:

✔ The **breakdown voltage,** commonly called the *Zener voltage,* is the reverse-biased voltage that causes the diode to break down and conduct current. Breakdown voltages, which are controlled by the semiconductor doping process, range from 2.4 V to hundreds of volts.

✔ The **power rating** tells you the maximum power (voltage × current) the Zener diode can handle. (Even diodes designed to break down can *really* break down if you exceed their power ratings.)

The circuit symbol for a Zener diode is shown here.

Because Zener diodes are so good at maintaining a constant reverse-biased voltage, even as current varies, they're used to regulate voltage in circuits. In the circuit in Figure 6-8, for example, a 9 V DC supply is being used to power a load, and a Zener diode is placed so that the DC supply exceeds the breakdown voltage of 6.8 V. (Note that this voltage is reverse-biasing the diode.) Because the load is in parallel with the Zener diode, the voltage drop across the load is same as the Zener voltage, which is 6.8 V. The remaining supply voltage is dropped across the resistor (which is there to limit the current through the diode so the power rating is not exceeded).

Here's the important thing: If the supply voltage varies up or down around its nominal 9 V value, the current in the circuit will fluctuate *but the voltage across the load will remain the same:* a constant 6.8 V. The Zener diode allows current fluctuations while stabilizing the voltage, whereas the resistor voltage varies as the current fluctuates.

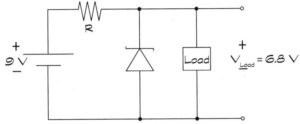

Figure 6-8: The Zener diode stabilizes the voltage drop across the load in this circuit.

Seeing the light with LEDs

All diodes release energy in the form of light when forward-biased. The light released by standard silicon diodes is in the infrared range, which is not visible to the human eye. *Infrared light-emitting diodes (IR LEDs)* are commonly used in remote-control devices to send secret (well, okay, invisible) messages to other electronic devices, such as your TV or DVD player.

Diodes known as *visible LEDs* (or just *LEDs*) are specially made to emit copious amounts of visible light. By varying the semiconductor materials used, diodes can be engineered to emit red, yellow, or green light, and there are special-purpose LEDs that emit blue or even white light. Bi-color and tri-color LEDs contain two or three different diodes within one package.

Here's the circuit symbol for an LED. The diode in an LED is housed in a plastic bulb designed to focus the light in a particular direction. The lead connected to the cathode is shorter than the lead connected to the anode. Compared to standard incandescent light bulbs, LEDs are more durable and efficient, run cooler, achieve full brightness much faster, and last much longer. LEDs are commonly used as indicator lights in automobiles, computers, and audio electronics, and are used in digital clocks and other displays. Figure 6-9 shows a single-color LED.

Figure 6-9:
The shorter
lead of
a typical
single-color
LED is
attached to
the cathode.

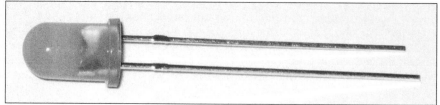

LEDs carry the same specifications as standard diodes, but they usually have fairly low current and PRV ratings. A typical LED has a PRV rating of about 5 V with a maximum current rating of under 50 mA. If more current passes through an LED than its maximum rating specifies, the LED burns up like a marshmallow in a campfire. Forward voltages vary, depending on the type of LED; they range from 1.5 V for IR LEDs up to 4.6 V for blue LEDs. Red, yellow, and green LEDs typically have a forward voltage of about 2.0 V. Be sure to check the specifications of any LEDs you use in circuits.

The maximum current rating for an LED is usually referred to as the maximum *forward current,* which is different from another LED rating, known as the *peak current* or *pulse current.* The peak/pulse current, which is higher than the maximum forward current, is the absolute maximum current that you can pass through the LED for a very short period of time. Here, short means *short* — on the order of milliseconds. If you confuse forward current with peak current, you may wreck your LED.

You should never connect an LED directly to a power source, or you may fry the LED instantly. Instead, use a resistor in series with the LED to limit the forward current. For instance, in the circuit in Figure 6-10, a 6 V battery is used to power a red LED. The LED has a forward voltage drop of 2.0 V and a maximum current rating of 30 mA. The voltage drop across the resistor is the difference between the source voltage and the LED forward voltage, or 6 V – 2 V = 4 V. The question is, how big should the resistor be to limit the current to 30 mA (that's 0.030 A) *or less* when the voltage dropped across the resistor is 4 V? You apply Ohm's Law (which we discuss in Chapter 3) to calculate the *minimum* value of resistance required to keep the current below the maximum current rating as follows:

$$R = \frac{V_R}{I_{max}}$$

$$= \frac{4\ V}{0.030\ A}$$

$$\approx 133\ \Omega$$

Chances are you won't find a resistor with the exact value you calculated, so choose a standard resistor with a *higher* value (such as 150 Ω) to limit the current a bit more. If you choose a lower value (such as 120 Ω), the current will exceed the maximum current rating.

Figure 6-10:
Be sure
to insert a
resistor in
series with
an LED to
limit current
to the LED.

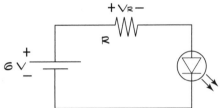

Other uses of diodes

Among the many other uses of diodes in electronic circuits are the following:

- **Overvoltage protection:** Diodes placed in parallel with a piece of sensitive electronic equipment protect the equipment from large voltage spikes. The diode is placed "backward" so that it's normally reverse-biased, acting like an open circuit and not playing any part in the normal operation of the circuit. However, under abnormal circuit conditions, if a large voltage spike occurs, the diode becomes forward-biased — which limits the voltage across the sensitive component and shunts excess current to ground to prevent harm to the component. (The diode may not be so lucky.)

- **Construction of logic gates:** Diodes are the building blocks of specialized circuits known as *logic,* which process signals consisting solely of two voltage levels that are used to represent binary information (such as on/off, high/low, or 1/0) in digital systems. We discuss logic a bit more in Chapter 7.

- **Current steering:** Diodes are sometimes used in uninterruptible power supplies (UPSs) to prevent current from being drawn out of a backup battery under normal circumstances, while allowing current to be drawn from the battery during a power outage.

Tremendously Talented Transistors

Imagine the world without the amazing electronics building block known as the transistor. Your cellphone would be the size of a washing machine, your laptop wouldn't fit on your lap (or in a single room), and your iPod would still be a gleam in Steve Jobs's eye.

Transistors are the heart of nearly every electronic device in the world, quietly working away without taking up much space, generating a lot of heat, or breaking down every so often. Generally regarded as the most important technological innovation of the 20th century, transistors were developed as an alternative to the vacuum tube, which drove the development of electronic systems ranging from radio broadcasting to computers, but exhibited some undesirable characteristics. The solid-state transistor enabled the miniaturization of electronics, leading to the development of cellphones, iPods, GPS systems — and much more.

Transistors in most portable gadgets these days are microscopically small, have no moving parts, are very reliable, and dissipate a heck of a lot less power than their vacuum-tube predecessors. They basically do just two things in electronic circuits: switch and amplify. But those two jobs are the key to getting things done. If you can switch electron flow on and off, you have control over the flow, and you can build very involved circuits by incorporating lots of switches in the right places. If you can amplify an electrical signal, then you can store and transmit tiny signals and boost them when you need them to make something happen (say, move the diaphragm of a speaker).

There are many different types of transistors The two most common types of transistors are

- Bipolar junction transistors
- Field-effect transistors

The sections that follow provide a closer look at these.

Bipolar junction transistors

The first transistors to be invented were *bipolar junction transistors* (BJTs), and BJTs are what most hobbyists use in homebrewed circuits. BJTs consist of two pn-junctions fused together to form a three-layer sandwich-like structure. Leads are attached to each section of the transistor, and are labeled the *base, collector,* and *emitter.* There are two types of bipolar transistors:

- **NPN transistors:** A thin piece of P-type semiconductor is sandwiched between two thicker pieces of N-type semiconductor, and leads are attached to each of the three sections. The symbol for an NPN transistor is shown here.

- **PNP transistors:** A thin piece of N-type semiconductor is sandwiched between two thicker pieces of P-type semiconductor, and leads are attached to each section. The symbol for a PNP transistor is shown here.

Bipolar transistors essentially contain two pn-junctions: the base-emitter junction and the base-collector junction. By controlling the voltage applied to the base-emitter junction, you control how that junction is biased (forward or reverse), ultimately controlling the flow of electrical current through the transistor. We explore exactly how an NPN transistor works in the section "How Transistors Really Work," later in this chapter.

Field-effect transistors

A *field-effect transistor (FET)* consists of a channel of N- or P-type semiconductor material through which current can flow, with a different material (laid across a section of the channel) that controls the conductivity of the channel. One end of the channel is known as the *source,* the other end of the channel is called the *drain,* and the control mechanism is called the *gate.* By applying a voltage to the gate, you control the flow of current from the source to the drain. Leads are attached to the source, drain, and gate. Some FETs include a fourth lead so you can ground part of the FET to the chassis of the circuit.(But don't confuse these four-legged creatures with *dual-gate MOSFETs,* which also have four leads.)

FETs come in two flavors — N-channel and P-channel — depending on the type of semiconductor material (N-type or P-type, respectively) through which current flows. There are two major sub-types of FETs: *MOSFETs (metal-oxide-semiconductor FETs)* and *JFETs (junction FETs).* Which is which depends on exactly how the gate is constructed — which results, in turn, in different electrical properties and different uses for each type. The details of gate construction are beyond the scope of this book, but you should be aware of the names of the two major types of FETs.

FETs (particularly MOSFETs) have become much more popular than bipolar transistors for use in integrated circuits, which we discuss in Chapter 7, where thousands of transistors work together to perform a task. That's because they're low-power devices whose structure allows thousands of N- and P-channel MOSFETs to be crammed together like sardines on a single piece of silicon.

Electrostatic discharge (ESD) can damage FETs. If you purchase FETs, be sure to keep them in an antistatic bag or tube — and leave them there until you're ready to use them. You can read more about the harmful effects of ESD in Chapter 9.

Operating a Transistor

BJTs and FETs work basically the same way. The voltage you apply to the input (*base*, for a BJT, or *gate*, for an FET) determines whether or not current flows through the transistor (from collector to emitter, for a BJT, and from source to drain for an FET).

- Below a certain voltage level, no current flows.
- Above a certain voltage level, the maximum current possible flows.
- In between those two voltage levels, an in-between amount of current flows.

In the "in-between" operating mode, small fluctuations in input current produce large fluctuations in output current. By allowing only the first two possibilities of input voltage (all or nothin'), you use the transistor as an on/off switch for current flow. By allowing the third possibility, you use the transistor as an amplifier.

To understand how a transistor works (specifically, a FET), think of a pipe connecting a source of water to a drain with a controllable valve across a section of the pipe, as shown in Figure 6-11. By controlling whether the valve is fully closed, fully open, or partially open, you control the flow of water from the source to the drain. You can set up the control mechanism for your valve in two different ways: It can act like an on/off switch, either fully opening or fully closing with nothing in between; or it can open partially, depending on how much force you exert on it. When it's partially open, you can adjust the valve a little to allow more or less water to flow from source to drain; small changes in the force you exert on the valve create similar, yet larger, changes in the flow of water. That's how a transistor acts as an amplifier.

Figure 6-11: In a field-effect transistor (FET), voltage applied to the gate controls the flow of current from the source to the drain.

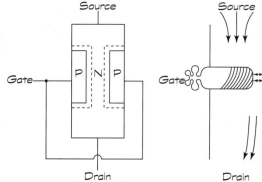

How Transistors Really Work

If you're curious about how applying a voltage to one pn-junction in a bipolar transistor can control the current flowing through the rest of the transistor, this section is for you. If you're not all that interested in the goings-on of free electrons and holes in doped semiconductors, you can skip this section all together and head off to the section, "Using a Model to Understand Transistors."

To get an insider's view of how transistors really work, we take a look at an NPN transistor. Figure 6-12 shows a representation of the structure of an NPN transistor, which includes a narrow P-type section sandwiched between two thicker N-type sections. The P-type section forms the base of the transistor, and has a lead sticking out of it. One of the N-type sections is the emitter and the other is the collector. The emitter and collector are not interchangeable — they're doped differently, so each has a different concentration of free electrons. Because the base is so narrow, there are a heck of a lot fewer holes available in the base than there are free electrons in the emitter and collector. This difference is important.

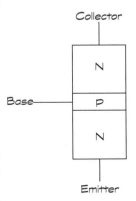

Figure 6-12:
The base of an NPN transistor sandwiches P-type semiconductor between two thicker N-type semiconductors.

Emitting and collecting electrons

There are two pn-junctions in an NPN transistor: the *base-collector junction* (the pn-junction between the base and the collector) and the *base-emitter junction* (the pn-junction between the base and the emitter). This is like sticking two diodes together, anode-to-anode. Say you connect two different voltage sources across the two junctions, as shown in Figure 6-13: One source, V_{CE}, applies a positive DC voltage from the collector to the emitter; the other source, V_{BE}, applies a positive DC voltage from the base to the emitter. If $V_{CE} > V_{BE}$, then the voltage at the base is more negative than the voltage at the collector, so the base-collector junction is reverse-biased, and no current flows across that junction. If you raise V_{BE} to about 0.7 V (the forward voltage of a silicon pn-junction) or more, you forward-bias the base-emitter junction — and current flows across that junction.

The current that flows across the base-emitter junction consists of the set of free electrons in the emitter — in effect, the emitter is sending ("emitting") its electrons toward the base. (Of course, the free electrons in the emitter are being pushed by an external force, which comes from the power-supply voltage connected to the collector.)

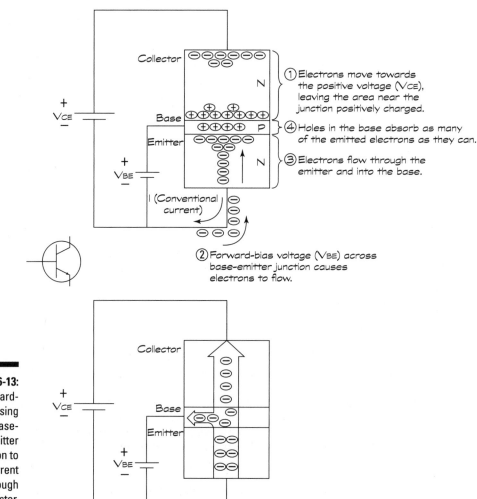

Figure 6-13:
Forward-
biasing
the base-
emitter
junction to
let current
flow through
a transistor.

Some of the electrons that get into the base *recombine* with the holes floating around in the base (remember, the base is made of a P-type semiconductor). But because the base is so narrow, there aren't enough holes to absorb all the electrons crossing the junction, so the base becomes negatively charged and tries to force the extra electrons out. There are two paths for all those surplus electrons to travel out of the base:

- ✔ Through the base connection that leads to the positive voltage source.
- ✔ Through the base-collector junction and into the collector.

So which way do they go? Remember that the base-collector junction is reverse-biased, with a strong positive voltage connected to the collector lead. That positive voltage tends to attract the free electrons that normally exist in the N-type collector toward one end of the collector. This makes electrons scarcer in the region at the *other* end of the collector (right near the base-collector junction) — so that end looks like it's positively charged.

Hmmm ... so the region of the collector right near the base-collector junction looks positively charged, and a whole slew of electrons that were emitted by the emitter are now crowding into the base looking for a way out. So what happens? *Most of those electrons (about 99 percent) get pulled across the base-collector junction.* Result: The electrons that were emitted by the emitter are now "collected" by the collector. A small percentage (less than 1 percent) of the electrons move out of the base through the lead that is connected to the base voltage source, but the pull across the base and out is not as strong as the other pull across the junction and into the collector. The collector wins the tug-of-war, attracting most of the emitter's electrons.

What you see as an outside observer, looking at the leads connected to the transistor, is that when the base-emitter junction is forward-biased, electrons flow from the emitter and then get split between the collector and the base, with most of the electrons (about 99 percent) going to the collector. *By controlling the voltage at the base-emitter junction, you cause a large amount of electrons to flow through the transistor from emitter to collector.* This is the crux of transistor operation. Biasing the transistor is like opening a valve that controls the flow of current through the transistor.

Now, electronic circuits always refer to conventional current — which, as mentioned earlier, is just the opposite of the real electron flow we've been talking about in this section. So, in circuit-speak, you say that forward-biasing the base-emitter junction of an NPN transistor causes a small (conventional) current to flow from the base to the emitter, and a large (conventional) current to flow from the collector to the emitter. That's why the circuit symbol for an NPN transistor shows an arrow pointing out of the emitter; it's pointing in the direction of *conventional* current flow.

You forward-bias a silicon NPN transistor by applying a voltage of at least 0.7 V to the base-emitter junction. Transistors made of germanium (much less common than silicon) have smaller forward-biased voltages (about 0.3 V), but the same cause-and-effect transistor action occurs. PNP transistors work the same way, except that all of the polarities are reversed because the junctions are reversed. To forward-bias the base-emitter junction of a PNP transistor, for example, you apply a voltage of –0.7 V from base to emitter. The circuit symbol for a PNP transistor shows an arrow pointing *into* the emitter, which indicates the direction of conventional current flow when the transistor is conducting.

Gaining current

When a transistor is conducting current, if you increase the current flowing into the base, I_B, an interesting thing happens: The current flowing into the collector, I_C, increases, too. If you decrease the base current, the collector current decreases, too. In fact, the 1%–99% relationship between the base current and the collector current holds true as you vary the base current (within limits, which we discuss in the section, "Saturating the transistor," later in this chapter).

The pattern of current changes at the collector exactly tracks the pattern of current changes at the base — but it's much bigger. This is why transistors are known as current amplifiers — even though they don't actually *create* larger currents; they just *control* them. The amount of collector current that flows is directly proportional to the amount of base current. The *current gain* (symbolized by h_{FE}) of the transistor depends on several factors, including the specific transistor selected.

Even for a single, specific transistor, the current gain varies depending on several factors. You should never design a circuit that hinges on a specific value of current gain, or your circuit is likely to act wacky. If you're using the transistor as an on/off switch (which we discuss in the section "Switching Signals with a Transistor"), it doesn't matter so much what the exact current gain is. If you're using the transistor as an amplifier, you can avoid a wacky circuit by configuring your transistor along with other components (such as resistors) in a way that removes any dependency on an exact value of current gain. It turns out this clever little workaround (which we discuss in the section "Amplifying Signals with a Transistor," later in this chapter) is very easy to design.

Saturating the transistor

In transistors, proportional current gain from base to collector occurs up to a point. Remember that the insides of a transistor consist of doped semiconductors that have a limited number of free electrons or holes that can move around. As you increase the voltage supply feeding into the base, you allow more electrons to flow out of the base, which means that more electrons are coming from the emitter. But there are only so many free electrons available in the emitter, so there's an upper limit to how much current can flow. When the transistor maxes out, it's *saturated*.

Understanding the word "transistor"

So why are transistors called transistors? Well, the word "transistor" is a combination of two word parts: *trans* and re*sistor.* The *trans* part of the name conveys the fact that by placing a forward-biased voltage on the base-emitter junction, you cause electrons to flow in another part of the component, from emitter to collector. You *transfer* the action from one part of the component to another. This is known as *transistor action.*

Because fluctuations in base current result in proportional fluctuations in collector/emitter current, you can think of the transistor as a sort of variable resistor: When you turn the dial (by varying the base current), the resistance changes, producing a proportionally varying collector/emitter current. That's where the *sistor* part of the name comes from.

You can think of saturating a transistor as opening a valve wider and wider, so that more and more water flows through a pipe, until the pipe is handling as much water as it can; even if you can open the valve more, you can't get any more water to flow through the pipe.

When a transistor is saturated, both of its junctions (base-emitter and base-collector) are forward-biased. The voltage drop across the output of the transistor — from collector to emitter — is nearly zero. It's as if there is a wire connected across the output of the transistor. And since I_C is so much larger than I_B, and $I_E = I_B + I_C$, you can say that $I_C \approx I_E$. You'll find that this approximation comes in handy when analyzing and designing transistor circuits.

If you operate the transistor so that the current either maxes out or doesn't flow at all, you're using the transistor as an on/off switch. You do this by designing your circuit so that the base-emitter junction is either nonconducting (the voltage across it is less than 0.7 V) or fully conducting — with nothing in between.

Using a Model to Understand Transistors

Free electrons, moving holes, pn-junctions, and biasing are all very nice, but you really don't have to know all that technical stuff by chapter and verse in order to use transistors in circuits. Instead, you can familiarize yourself with a functional model of a transistor, and you'll know enough to get going.

Figure 6-14 shows a simple model of an NPN transistor on the left, and the circuit symbol for an NPN transistor on the right. Voltages, currents, and transistor terminals are labeled for both so you can see how the model corresponds to the actual device. The model includes a diode between the base and the emitter controlling a variable resistance, R_{CE}, between the collector and the emitter.

Choosing the right switch

You may wonder why you would use a transistor as a switch when there are so many other types of switches and relays available (as Chapter 8 describes). Well, transistors have several advantages over other types of switches, and so are used when they're the best choice. Transistors use very little power, can switch on and off several billion times per second, and can be made microscopically small, so

integrated circuits (which we discuss in Chapter 7) use thousands of transistors to switch signals around on a single tiny chip. Mechanical switches and relays have their uses, too, in situations where transistors just can't handle the load, such as switching currents bigger than about 5 A or switching higher voltages (as in electrical power systems).

There are three different *operating modes,* or possibilities for how the transistor operates:

- **Transistor off:** If V_{BE} < 0.7 V, the diode is off, so I_B = 0. This makes the resistance R_{CE} infinite, which means I_C = 0. The output of the transistor (collector-to-emitter) is like an open switch: No current is flowing. You call this mode of operation *cutoff.*

- **Transistor partially on:** If $V_{BE} \geq 0.7$ V, the diode is on, so base current flows. If I_B is small, the resistance R_{CE} is reduced and some collector current, I_C, flows. I_C is directly proportional to I_B, with a *current gain,* h_{FE}, equal to I_C/I_B, and the transistor is functioning as a current amplifier — that is, operating in *active* mode.

- **Transistor fully on:** If $V_{BE} \geq 0.7$ V, and I_B is increased a lot, the resistance R_{CE} is zero and the maximum possible collector current, I_C, flows. The voltage from collector to emitter, V_{CE}, is nearly zero, so the output of the transistor (collector-to-emitter) is like a closed switch: All current that can flow through it is flowing; the transistor is *saturated.*

Figure 6-14:
A transistor works as a switch or an amplifier, depending on what you input to the base.

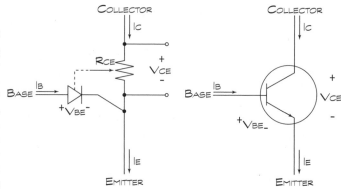

When you design a transistor circuit, you choose components that will put the transistor into the right operating mode (cutoff, active, or saturation), depending on what you want the transistor to do. For instance, if you want to use the transistor as an amplifier, you select supply voltages and resistors that will forward-bias the base-emitter junction and allow just enough base current to flow — but not so much that the transistor becomes saturated. This selection process is known as *biasing* the transistor. If you want the transistor to act like a switch, you choose values that allow only two states — either no base current flows at all, or enough base current flows to saturate the transistor — and you use either a switch or the output of a previous stage of electronic circuitry to control that two-state operation.

Amplifying Signals with a Transistor

Transistors are commonly used to amplify small signals. Say, for instance, you produce an audio signal as the output of one stage of an electronic circuit, and you'd like to amplify it before shipping it off to another stage of electronics, such as a speaker. You use a transistor, as shown in Figure 6-15, to amplify the small up-and-down fluctuations in the audio signal (v_{in}), which you input to the base of the transistor; there they become *large* signal fluctuations (v_{out}), which appear at the output (collector) of the transistor. Then you take the transistor output and apply it to the input of your speakers.

Figure 6-15:
By strategically positioning a few resistors in a transistor circuit, you can properly bias a transistor and control the gain of the circuit.

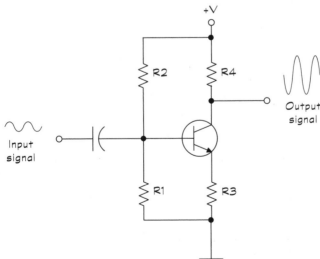
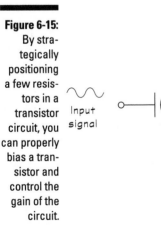

Biasing the transistor so it acts like an amplifier

A transistor must be partially on in order to work as an amplifier. To put a transistor in this state, you bias it by applying a small voltage to the base. In the example in Figure 6-15, resistors R1 and R2 are connected to the base of the transistor and configured as a voltage divider (for more about how a voltage divider works, see Chapter 3). The output of this voltage divider ($\frac{R1}{R1+R2} \times V$) supplies enough voltage to the base to turn the transistor on and allow current to flow through it, biasing the transistor so it's in the active mode (that is, partially on).

The capacitor at the input allows only AC to pass through to the transistor, blocking any DC component of the input signal (an effect known as a *DC offset*), as shown in Figure 6-16. Without that capacitor, any DC offset in the input signal could upset the bias of the transistor, potentially shutting the transistor off (cutoff) or saturating it so it no longer acts like an amplifier.

Figure 6-16:
A blocking capacitor helps maintain the bias of the transistor by filtering out DC offsets in the input signal.

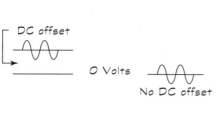

Controlling the voltage gain

With the transistor in Figure 6-15 partially on, the current fluctuations caused by the AC input signal get amplified by the transistor. Because the current gain of any transistor you happen to choose can be somewhat variable (schizophrenic, in fact), you design your amplifier circuit in such a way as to eliminate any dependency on the flaky current gain. You'll give up some strength of amplification, but you'll get stability and predictability in return.

By placing resistors R3 and R4 in the circuit, you can control the *voltage gain,* or how much the input signal is amplified — without worrying about the exact current gain of the specific transistor at the heart of your circuit. (This is truly amazing stuff!) The AC voltage gain of a transistor circuit with resistors as shown in Figure 6-15 is $-R_4/R_3$. The negative sign just means that the input signal is *inverted:* As the input voltage varies up and then down, the output voltage varies down and then up, as shown by the input and output signal waveforms in Figure 6-15.

Configuring transistor amplifier circuits

The type of transistor setup we discuss in the preceding section is known as a *common-emitter amplifier;* this circuit is just one of many different ways to configure transistor circuits for use as amplifiers. You use different configurations to achieve different goals, such as high power gain versus high voltage gain. How the circuit behaves depends on

- ✔ How you connect the transistor to the power supplies.
- ✔ Where the load is.
- ✔ What other circuit components (such as resistors, capacitors, and other transistors) you add to the circuit.
- ✔ Where you add other components in the circuit.

For instance, you can piggyback two bipolar transistors in a setup known as a *Darlington pair* to produce multiple stages of amplification. (In Chapter 14, you learn exactly how to configure a simple Darlington pair.) Or you can get the same result the easy way: Purchase a three-lead component called a *Darlington transistor,* which includes a Darlington pair already hooked up.

 Designing transistor amplifier circuits is a field of study on its own, and many excellent books have been written on the subject. If you're interested in learning more about transistors and how to design amplifier circuits using transistors, try getting your hands on a good electronics design book, such as *The Art of Electronics,* by Thomas C. Hayes and Paul Horowitz (Cambridge University Press). It isn't cheap, but it's a classic.

Switching Signals with a Transistor

You can also use a transistor as an electrically operated switch. The base lead of the transistor works like the toggle on a mechanical switch. The transistor is "off" when no current flows into the base (in cutoff), and the transistor acts like an open circuit — even if there is a voltage difference from collector to emitter. The transistor is "on" when current flows into the base (in saturation), and the transistor acts like a closed switch, allowing current to flow from collector to emitter — and out to whatever load you want to turn on.

How do you get this on-off thing to work? Say you use an electronic gadget to scatter chicken feed automatically at dawn. You can use a *photodiode,* which conducts current when exposed to light, to control the input to a transistor switch that delivers current to your gadget (the load). At night, the photodiode doesn't generate any current, so the transistor is off. When the sun rises, the photodiode generates current, turning the transistor on and allowing current to flow to your gadget. The gadget then starts scattering chicken feed — keeping the chickens happy while you continue to snooze.

If you're wondering why you don't just supply the current from the photodiode to the gadget, it's because your gadget might need a larger current than can be supplied by the photodiode. The small photodiode current controls the on/off action of the transistor, which acts like a switch to allow a larger current from a battery to power your gadget.

One of the reasons transistors are so popular for switching is that they don't dissipate a lot of power. Remember that power is the product of current and voltage. When a transistor is off, no current flows, so the power dissipated is zero. When a transistor is fully on, V_{CE} is nearly zero, so the power dissipated is nearly zero.

Choosing Transistors

Transistors have become so popular, there are literally thousands upon thousands of different transistors currently available. So how do you choose one for your circuit, and how do you make sense out of all the choices available on the market?

If you're designing a transistor circuit, you need to understand how your circuit will operate under various conditions. What is the maximum amount of collector current your transistor will have to handle? What is the minimum current gain you need in order to amplify an input signal? How much power could possibly be dissipated in your transistor under extreme operating conditions (for instance, when the transistor is off, and the entire power supply voltage may be dropped across the collector-emitter)?

After you understand the ins-and-outs of how your circuit will operate, you can start looking up transistor specifications to find one that meets your needs.

Important transistor ratings

There are loads of parameters used to describe the loads of different transistors available on the market, but only a few you really need to know in order to choose the right transistor for your circuit. For bipolar (NPN or PNP) transistors, here's what you need to know:

- **I_C max:** This is the *maximum collector current* the transistor can handle. When designing a circuit, make sure you use a resistor to limit the collector current so it doesn't exceed this value.

- **h_{FE}:** This is the *DC current gain* from base to collector. Because the current gain can vary — even among transistors of the same type — you need to know the guaranteed minimum value of h_{FE}, and that's what this parameter tells you. The h_{FE} also varies for different values of I_C, so sometimes h_{FE} is given for a specific value of I_C, such as 20 mA.

- **V_{CE}:** This is the *maximum collector-to-emitter voltage.* It's usually at least 30 V. If you're working with low-power applications such as hobby electronics circuits, don't worry about this value.

- **P_{total} max:** This is the *maximum power dissipation,* which is roughly $V_{CE} \times I_C$ max. No need to worry about this rating if you're using the transistor as a switch; power dissipation is nearly zero anyway. If you're using the transistor as an amplifier, however, you need to be aware of this rating.

If you think your circuit will approach this value, be sure you attach a heat sink to the transistor.

Of course, none of these ratings appears anywhere on the transistor itself — that would be too easy. To determine these characteristics, you have to look up the transistor in a specifications book, or consult the technical documentation at the manufacturer's Web site. If you're building a circuit someone else designed, you can simply use the transistor specified by the designer, or consult a cross-reference book to find a similar model to substitute.

Identifying transistors

Many bipolar transistors originating in North America are identified by a five- or six-digit code that is part of an industry-standard semiconductor identification system. The first two digits are always "2N" for transistors, with the "2" specifying the number of pn-junctions, and the "N" signifying a semiconductor. The remaining three or four digits indicate the specific features of the transistor. However, different manufacturers may use different coding schemes, so your best bet is to consult the appropriate Web site, catalog, or specification sheet to make sure you're getting what you need for your circuit.

Many suppliers categorize transistors according to the type of application they are used in, such as low-power, medium-power, high-power, audio (low-noise), or general-purpose. Knowing the category that describes your project can help guide you to make the right selection of a transistor for your particular circuit.

Recognizing a transistor when you see one

The semiconductor material in a transistor is the size of a grain of sand or even smaller, so manufacturers put these teensy little components in a metal or plastic case with leads sticking out so you can connect them in your circuits. You can find literally dozens upon dozens of different shapes and sizes of transistors, some of which are shown in Figure 6-17. The smaller packages generally house *signal transistors,* which are rated to handle smaller currents, while larger packages contain *power transistors* designed to handle larger currents. Most signal transistors come in plastic cases, but some precision applications require signal transistors housed in metal cases to reduce the likelihood of stray radio-frequency (RF) interference.

Figure 6-17:
Signal and power transistors come in a variety of sizes and packages.

Bipolar transistors typically have three wire leads so you can access the base, collector, and emitter of the transistor. One exception to this is a *phototransistor* (which we discuss in Chapter 8), which is packaged in a clear case and has just two leads (collector and emitter), because light is used to bias the transistor so you don't have to apply a voltage to the base. All FETs have leads for the source, drain, and gate, and some include a fourth lead so you can ground the transistor's case to the chassis of your circuit, or for the second gate of a dual-gate MOSFET.

To figure out which package lead is which, consult the documentation for the specific transistor. Be careful how you interpret the documentation: Transistor connections are often (though not always) shown from the underside of the case, as if you've turned the transistor over and are gazing at it from the bottom.

You absolutely must install transistors the proper way in your circuits. Switching the connections around can damage a transistor and may even damage other circuit components.

Making All Kinds of Components Possible

Transistors can be combined in all sorts of different ways to make lots of incredible things happen; because the actual semiconductor material that makes up a transistor is so small, it's possible to create a circuit containing hundreds or thousands of transistors (along with resistors and other components) and plop the entire circuit into a single component that fits easily into the palm of your hand. These amazing creations, known as *integrated circuits (ICs),* enable you to build *really* complex circuits with just a couple of parts. In the next chapter, we'll take a look at some of the ICs that are available today as a result of the semiconductor revolution.

Trying Your Hand at Semiconductor Circuits

If you want to switch gears and gain firsthand experience with diodes and transistors, check out the learning circuits in the middle of Chapter 14 in Part III. There, you'll find simple circuits designed to show you how these components work. You can turn an LED on and off, and vary the intensity of the light by controlling the amount of current passing through it. You can observe a Zener diode holding an output voltage steady. Simple transistor circuits show you how transistors amplify current and switch current on and off. Before you power up a semiconductor device, be sure to read through the safety precautions in Chapter 9 and review the other chapters in Part II to find out how to build and analyze circuits. That way, you'll get maximum gain out of your experience with semiconductors.

Chapter 7

Packing Parts Together on Integrated Circuits

· ·

In This Chapter

▶ Corralling components into a chip

▶ Speaking the language of bits

▶ Thinking logically about gates

▶ Reading into IC packages

▶ Pondering IC pinouts

▶ Boosting signals with op amps

▶ Timing, counting, and controlling everything in sight

· ·

Space exploration, programmable pacemakers, consumer electronics, and much more would be nothing but the idle dreams of creative minds were it not for the integrated circuit. This incredible innovation — really, a series of incredible innovations — makes possible your cellphone, laptop, iPod, GPS navigation system, and so much more.

An *integrated circuit* (IC) incorporates anywhere from a few dozen to many millions of circuit components into a single device that fits easily into the palm of your hand. Each IC contains an intricate mesh of tiny transistor-based taskmasters, with access to the outside world provided via a finite number of inputs and outputs.

This chapter explores how integrated circuits came to be, identifies the three major IC flavors, and dissects the inner workings of one variety — digital ICs. You get a look at how computers and other digital devices manipulate two distinct voltage levels to process information using special rules known as logic. Next up is an explanation of how to "read" an IC to understand what the heck it does (because you can't tell by its cover) and how to connect it up for use in real circuits. Finally, you get a closer look at three best-selling ICs, what they do, and how you can use them to create your own innovative circuits.

Why ICs?

The integrated circuit (IC) was invented in 1958 (see the sidebar, "The birth of the IC") to solve the problems inherent in manually assembling mass quantities of tiny transistors. Also called *chips,* integrated circuits are miniaturized circuits produced on a single piece of semiconductor. A typical integrated circuit contains hundreds of transistors, resistors, diodes, and capacitors; the most advanced ICs contain several hundred million components. Because of this circuit efficiency, you can build really complex circuits with just a couple of parts. ICs are the building blocks of larger circuits. You string them together to form just about any electronic device you can think up.

The birth of the IC

With the invention of the transistor in 1947, the focus of electronic design shifted away from bulky vacuum tubes to this newer, smaller, more reliable device. This stirred up tremendous excitement, as engineers worked to build more and more advanced circuits because size was no longer an obstacle. Their success in creating advanced designs led to some practical problems: Interconnecting hundreds of components inevitably resulted in errors that were extremely difficult to isolate. Additionally, complex circuits often failed to meet speed requirements (because it does take some time for electrons to travel through a maze of wires and components). Throughout the 1950s, a major focus within the electronics industry was figuring out how to make circuits smaller and more reliable.

In 1952, a British engineer named Geoffrey Dummer publicly presented his idea for combining multiple circuit elements onto a single piece of semiconductor material with no connecting wires. He reasoned this would eliminate the faulty wiring and the cumbersome manual assembly of discrete components. Although Dummer never actually built an IC, he is widely regarded as "The Prophet of the Integrated Circuit."

Then, in the summer of 1958, Jack Kilby, a newly employed engineer at Texas Instruments working alone in a lab (while his colleagues were on vacation) was able to build multiple circuit components out of a single, monolithic piece of germanium (a semiconductor material), and lay metal connectors in patterns on top of it. Kilby's crude design was the first successful demonstration of the integrated circuit. Six months later, Robert Noyce of Fairchild Semiconductor (who also co-founded Intel) invented his own version of the IC, which solved many of the practical problems inherent in Kilby's design and led to the mass production of ICs. Together, Kilby and Noyce are credited with the invention of the integrated circuit. (Kilby was awarded the Nobel Prize in Physics for his contributions to the invention of the integrated circuit — but not until 42 years later — and stated that had Noyce been alive when the Prize was awarded, he surely would have shared it.)

A lot has happened since 1958. All those really smart, creative people continued to plug away at their work, and many more innovations took place. As a result, the electronics industry has exploded as *chip densities* (a measure of how closely packed the transistors are) have increased exponentially. Today, semiconductor manufacturers routinely carve millions of transistors into a piece of silicon smaller than the size of a dime. (Kinda makes your head spin, doesn't it?)

Linear, Digital, or Combination Plate?

Over the years, chip makers have come out with lots of different ICs, each of which performs a specific function depending on how the components inside are wired up. Many of the integrated circuits you encounter are so popular they have become standardized, and you can find a wealth of information about them online and in books. A lot of different chip makers offer these standardized ICs, and manufacturers and hobbyists the world over buy and use them in various projects. Other so-called special-purpose ICs are designed to accomplish a unique task. More often than not, only a single company sells a particular special-purpose chip.

Whether standardized or special-purpose, you can separate ICs into three main categories: *linear (analog), digital, and mixed signal.* These terms relate to the kinds of electrical signals (more about those Chapter 2) that work within the circuit:

- **Linear (analog) ICs:** These ICs contain circuits that process *analog signals,* which consist of continuously varying voltages and currents. Such circuits are known as *analog circuits.* Examples of analog ICs are power-management circuits, sensors, amplifiers, and filters.

- **Digital ICs:** These ICs contain circuits that process *digital signals,* which are patterns consisting of just two voltage (or current) levels representing binary digital data, for instance, on/off, high/low, or 1/0. (We discuss digital data a bit more in the next section.) Such circuits are known as *digital circuits.* Some digital ICs, such as microprocessors, contain millions of tiny circuits within just a few square millimeters.

- **Mixed signal ICs:** These ICs contain a combination of analog and digital circuits.

The majority of standardized ICs fall into either the linear or digital category, and most mail order businesses that sell ICs separate them into linear and digital lists.

Making Decisions with Logic

When you first learned to add numbers, you memorized facts such as "2 + 2 = 4," "3 + 6 = 9," and so forth. Then, when you learned to add multi-digit numbers, you used those simple facts as well as a new one — "carrying" numbers to us older folks, "regrouping" numbers to the younger generation. By applying a few simple addition facts and one simple rule, you can add two large numbers together fairly easily.

Adding up bits of numbers

In the decimal (*base 10*) system, if you want to express a number greater than 9, you need to use more than one digit. Each position, or place in a decimal number represents a *power of ten* (10^0, 10^1, 10^2, 10^3, and so forth), and the value of the digit (0–9) sitting in that position is a multiplier for that power of ten. With powers of ten, the *exponent* (that tiny number raised up next to the 10) tells you how many times to multiply 10 times itself, so 10^1 equals 10, 10^2 equals 10×10 which is 100, 10^3 equals $10 \times 10 \times 10$ which is 1,000, and so on. As for 10^0, it just equals 1 because *any* number raised to the zeroth power equals 1. So the positions in a decimal number, starting from the rightmost position, represent 1, 10, 100, 1,000, and so forth. These are also known as *place values* (ones or units, tens, hundreds, thousands, and so forth). The digit (0–9) sitting in that position (or place) tells you how many ones, tens, hundreds, thousands, and so forth are contained in that decimal number.

For example, the number 9,452 can be written in *expanded notation* as:

$$(9 \times 1,000) + (4 \times 100) + (5 \times 10) + (2 \times 1)$$

Our entire mathematics system is based on the number 10 (but if humans only had eight fingers, we might be using a *base 8* system), so your brain has been trained to automatically think in decimal format (it's like a math language). When you add two digits together, such as 6 and 7, you automatically interpret the result, 13, as "1 group of 10 plus 3 groups of 1." It's ingrained in your brain, no less than your native language is.

Well, the binary system is like another language: It uses the exact same methodology, but it's based on the number 2. If you want to represent a number greater than 1, you need more than one digit and each position in your number represents a *power of two*: 2^0, 2^1, 2^2, 2^3, 2^4, and so forth, which is the same as 1, 2, 4, 8, 16, and so forth. The *bit* (a bit is a binary digit, just 0 or 1) that sits in that position in your number is a multiplier for that power of two. For example, the *binary number* 1,101 can be written in expanded notation as:

$$(1 \times 2^3) + (1 \times 2^2) + (0 \times 2^1) + (1 \times 2^0)$$

By translating this into decimal format, you can see what numerical quantity the bit string 1,101 represents:

$$(1 \times 8) + (1 \times 4) + (0 \times 2) + (1 \times 1)$$
$$= \quad 8 \quad + \quad 4 \quad + \quad 0 \quad + \quad 1$$
$$= 13 \text{ (in decimal format)}$$

So the binary number 1,101 is the same as the decimal number 13. They're just two different ways of representing the same physical quantity. This is analogous to saying "bonjour" or "buenos días" rather than "good day." They're just different words for the same greeting.

When you add two binary numbers together, you use the same methodology that is used in the decimal system, but using 2 as a base. In the decimal system, $1 + 1 = 2$, but in the binary system, $1 + 1 = 10$ (remember, the binary number 10 represents the same quantity as the decimal number 2). Computers use the binary system for arithmetic operations because the electronic circuits inside computers can work easily with bits, which are just high or low voltages (or currents) to them. The circuit that performs addition inside a computer contains several transistors arranged in just the right way so that when high or low signals representing the bits of two numbers are applied to the transistor inputs, the circuit produces the right combination of high or low outputs to represent the bits of the numerical sum. Exactly how this is done is beyond the scope of this book, but hopefully, you now have an idea of how this sort of thing works.

The microprocessor in your computer works in much the same way. It uses lots of teeny little digital circuits — known among computer types as *digital logic* — to process simple functions similar to "2 + 2 = 4." Then the logic combines the outputs of those functions by applying rules similar to carrying/regrouping to get an answer. By piggybacking lots of these "answers" together in a complex web of circuitry, the microprocessor can perform some pretty complicated mathematical tasks. Deep down inside, though, there's just a bunch of logic applying simple little rules.

In this section, we take a look at how digital logic circuits work.

Beginning with bits

When you add two digits together, you have ten choices for each digit (0 through 9) because that's how our numbering system (known as a *base 10* or *decimal system*) works. When a computer adds two digits together, it uses only two possible digits: 0 and 1 (this is known as a *base 2* or *binary system*). Because there are only two, these digits are known as <u>bi</u>nary dig<u>its</u>, or *bits*. Bits can be strung together to represent letters or numbers — for instance, the bit string 1101 represents the number 13. The accompanying sidebar offers a glimpse of how this works.

In addition to representing numbers and letters, bits can also be used to carry information. As information carriers, data bits are versatile: They can represent many *two-state* (binary) things: A screen's pixel is either on or off; a CTRL key is either up or down; a laser pit is either present or absent on a DVD surface; an ATM transaction is either authorized or not — and much more. By assigning logical values of 1 and 0 to a particular on/off choice, you can use bits to carry information about real, physical events — and allow that information to control other things by processing the bits in a digital circuit.

Logical 1 and logical 0 are also referred to as true and false, or high and low. But what exactly *are* these "ones and zeros" in a digital circuit? They are simply high or low currents or voltages that are controlled and processed by transistors. (In Chapter 6, we discuss how transistors work and how they can be used as on/off switches.) Common voltage levels used to represent digital data are 0 volts for logical 0 (low) and (often) 5 volts for logical 1 (high).

A *byte,* which you've probably heard about quite a bit, is a grouping of eight bits used as a basic unit of information for storage in computer systems. Computer memory, such as Random Access Memory (RAM), and storage devices such as CDs and memory sticks, use bytes to organize gobs of data. Just as banks pack 40 quarters into a quarter roll, 50 dimes into a dime roll, and so forth to simplify the process of supplying merchants with change for their cash registers, so computer systems pack data bits together in bytes to simplify the storage of information.

Processing data with gates

Logic gates, or simply *gates,* are tiny digital circuits that accept one or more data bits as inputs and produce a single output bit whose value (1 or 0) is based on a specific rule. Just as different arithmetic operators produce different outputs for the same two inputs (for instance, three *plus* two produces five, while three *minus* two produces one), so different types of logic gates produce different outputs for the same inputs:

- ✔ **AND gate:** The output is high (1) only if both inputs (one input AND the other input) are high. If either input is low (0), the output is low. A standard AND gate has two inputs, but you can also find three-, four-, and eight-input AND gates. For those gates, the output is high only if *all* inputs are high.

- ✔ **NAND gate:** This function behaves like an AND gate followed by an inverter (hence the NAND, which means NOT AND). It produces a low output only if all of its inputs are high. If any input is low, the output is high.

- ✔ **OR gate:** The output is high if one OR the other OR both of its inputs are high. It only produces a low output if both inputs are low. A standard OR gate has two inputs, but three- and four-input OR gates are also available. For these gates, a low output is generated only when all inputs are low; if one or more inputs is high, the output is high.

- ✔ **NOR gate:** This behaves like an OR gate followed by a NOT gate. It produces a low output if one or more of its inputs are high, and generates a high output only if all inputs are low.

- ✔ **XOR gate:** The exclusive OR gate produces a high output if either one input OR the other is high, but not both; otherwise it produces a low output. All XOR gates have two inputs, but multiple XOR gates can be cascaded together to create the effect of XORing multiple inputs.

- ✔ **XNOR gate:** The exclusive NOR gate produces a low output if either one input or the other is high, but not both. All XNOR gates have two inputs.

- ✔ **NOT gate (inverter):** This single-input gate produces an output that inverts the input: A high input generates a low output, and a low input generates a high output. A more common name for a NOT gate is an *inverter.*

Figure 7-1 shows the circuit symbols for these common logic gates.

Most logic gates are built using diodes and transistors, which we discuss in Chapter 6. Inside each logic gate is a circuit that arranges these components in just the right way, so that when you apply input voltages (or currents) representing a specific combination of input bits, you get an output voltage (or current) that represents the appropriate output bit. The circuitry is built into a single chip with leads, known as *pins,* providing access to the inputs, outputs, and power connections in the circuit.

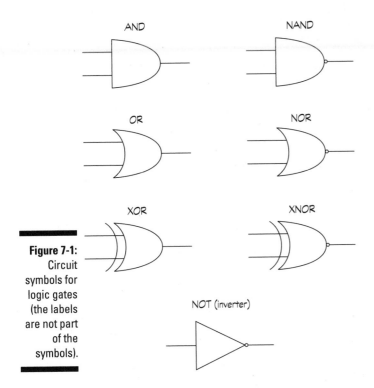

Figure 7-1:
Circuit symbols for logic gates (the labels are not part of the symbols).

You usually find multiple logic gates sold in integrated circuits, such as an IC containing four two-input AND gates (called a *quad 2-input* AND gate), as shown in Figure 7-2. The package sports pins that lead to each gate's inputs and output, as well as other pins that connect a power supply to the circuitry. Look on the Web site of the IC's manufacturer for a datasheet that tells you which pins are inputs, outputs, V+ (voltage), and ground. A *datasheet* is like a user's manual; it provides technical specifications and performance information about the chip.

Make sure that the part you buy has the number of inputs that you need for your project. Remember that you can buy logic gates with more than two inputs. For example, you can find a 3-input AND gate from most electronics suppliers.

By combining just NAND gates or just NOR gates in just the right way, you can create any of the other logical functions. Chip makers typically build digital circuits using NAND or NOR gates almost exclusively, so that they can focus their research and development efforts on improving the process and design of just two basic logic gates. That's why NAND and NOR gates are sometimes called *universal gates*.

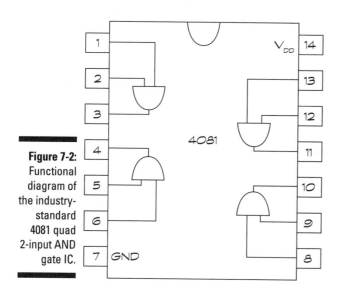

Figure 7-2:
Functional
diagram of
the industry-
standard
4081 quad
2-input AND
gate IC.

Simplifying gates with truth tables

Tracking all the high and low inputs to logic gates and the outputs they produce
can get a bit confusing — especially for gates with more than two inputs — so
designers use a tool called a *truth table* to keep things organized. This table lists
all the possible combinations of inputs and corresponding outputs for a given
logical function. Table 7-1 shows truth tables for the AND, NAND, OR, NOR, XOR,
XNOR, and NOT (inverter) logic gates: The first two columns, A and B, represent
input bits; the remaining columns show the output of the various gates.

Table 7-1		Truth Tables for Logic Gates						
A	*B*	*A AND B*	*A NAND B*	*A OR B*	*A NOR B*	*A XOR B*	*A XNOR B*	*NOT A*
0	0	0	1	0	1	0	1	1
0	1	0	1	1	0	1	0	1
1	0	0	1	1	0	1	0	0
1	1	1	0	1	0	0	1	0

You can also use truth tables for other digital circuits, such as a *half-adder* circuit, which is designed to add two bits and produce an output consisting of
a sum bit and a carry bit. For instance, for the binary equation 1 + 1 = 10, the
sum bit is 0 and the carry bit is 1. The truth table for the half adder is shown
below.

A	B	Carry	Sum
0	0	0	0
0	1	0	1
1	0	0	1
1	1	1	0

If you look at the carry-bit column in the truth table for the half-adder, you may notice that it looks just like the output for the two-input AND gate shown in Table 7-1: That is, the carry bit is the same as A AND B, where A and B are the two input bits. Similarly, the sum bit is the same as A XOR B. What's the significance of this? You can build a half-adder using an AND gate and an XOR gate. You feed the input bits into both gates, and use the AND gate to generate the carry bit and the XOR gate to generate the sum bit. (See Figure 7-3.)

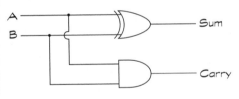

Figure 7-3:
The half-adder circuit consists of an AND gate and an XOR gate.

Creating logical components

By connecting several adders together in just the right way, you can create a larger digital circuit that takes two multi-bit inputs, such as 10110110 and 00110011, and produces their sum, 11101001. (In decimal notation, that sum is 182 + 51 = 233.)

You can create loads of other complex functions by combining multiple AND, OR, and NOT gates together. It's all a matter of which gates you use and how you interconnect them. Think about forming words from letters. With just 26 different choices, you can create millions of words. Likewise, you can create circuits that perform math functions (such as adders, multipliers, and many others) by connecting a whole bunch of gates together in the right combination.

Storing bits in registers

Connecting dozens of gates together in a complex web of circuitry presents a bit of a timing problem. As the gate inputs in one stage of logic change, the gate outputs will change — but not instantaneously (it takes some time, albeit very small, for each gate to react). Those outputs then feed the inputs to another stage of logic, and so on. Complex logic devices use special circuits called *registers* between stages of logic to hold (or store) the output bits of one stage for a short time before allowing them to be applied to the next logic stage. Registers ship their contents out and accept new contents in upon receiving a signal known as a *clock tick,* which gives every gate enough time to compute its output. Clock signals are produced by special precision timing circuits. (See the section on the *555 timer* later in this chapter for more information on how to create clocks and registers.)

Over the years, digital circuit designers have perfected the design of adders and other commonly used digital circuits, figuring out clever ways to speed up the computation time, reduce power dissipation, and ensure the results are accurate, even under harsh circuit conditions such as extreme temperatures. Tried-and-true digital circuit designs are commonly turned into standardized IC product offerings — so you and other circuit builders don't have to recreate the wheel over and over again.

Understanding How to Use ICs

Integrated circuits are nothing like discrete components — such as individual resistors, capacitors, and transistors — which have two or more leads connected directly to the component inside the package. The miniature pre-built components inside an IC are already interconnected into one big happy circuit, ready to perform a specific task. You just have to add a few ingredients — say, power and one or more input signals — and the IC will do its thing. Sounds simple, right? Well, it is. You just have to know how to "read" IC packages — because they all look like black multi-legged critters — so you know how to make the right connections.

Identifying ICs with part numbers

Every IC has a unique code, such as 7400 or 4017, to identify the type of device — really, the circuit — that's inside. You can use this code, also known as a part number, to look up specifications and parameters about an IC in a reference book or online resource. The code is printed on the top of the chip.

Many ICs also contain other information, including the manufacturer's catalog number and maybe even a code that represents when the chip was made. Don't confuse the date code or catalog number with the part number used to identify the device. Manufacturers don't have a universal standard for how they stamp the date code on their integrated circuits, so you may have to do some detective work to pick out the actual part number of the IC.

Packaging is everything

Great things really do come in small packages. Many ICs that can fit in the palm of your hand contain incredibly complex circuitry; for example, an entire AM/FM radio circuit (minus the battery and antenna) fits in an IC package the size of a nickel. The actual circuit is so small that manufacturers have to mount it onto a reasonable-size plastic or ceramic package so humans can use it. During the process of *chip assembly,* leads are attached to the appropriate circuit access points and fed out of the package enclosure so you and others like you can get current flowing to, through, and from the circuit inside.

Many ICs used in hobby electronics projects are assembled into *dual in-line packages (DIPs),* such as the ones in Figure 7-4. DIPs (sometimes called DILs) are rectangular-shaped plastic or ceramic packages with two parallel rows of leads, called *pins,* down either side. DIPs contain between 8 and 52 pins, but the most common sizes are 8-, 14-, and 16-pin. DIPs are designed to be *through-hole mounted* onto a printed circuit board (PCB), with the pins extending through holes in the board and soldered on the other side. You can solder DIP pins directly onto a circuit board or use *sockets* designed to hold the chip without bending the pins. You solder the socket connections into your circuit, and then insert the chip into the socket. DIPs also fit nicely into the contact holes of solderless breadboards (which we cover in Chapter 11), making it easy to prototype a circuit.

ICs used in mass produced products are generally more complex and require a higher number of pins than DIPs can provide, so manufacturers have developed (and continue to develop) clever ways of packaging ICs and connecting them to printed circuit boards (PCBs). To save space on the board (known as *real estate*), most ICs today are mounted directly to metal connections built onto the PCBs. This is known as *surface-mount technology (SMT),* and many IC packages are specially designed to be used this way. One such surface-mount IC package is the *small-outline integrated circuit (SOIC),* which looks like a shorter, narrower DIP with bent leads (called "gull wing" leads).

SMT packaging has become so widely adopted, that it is often difficult to find certain ICs sold in a DIP package. If you want to use a surface-mount IC in a solderless breadboard (because you may not be able to find the DIP variety), look for special DIP adapter modules that convert various surface-mount IC packages to pin compatible DIP packages you can plug directly into your breadboard. (Enter "DIP adapter" into your favorite Internet search engine to get a list of suppliers of such devices.)

Figure 7-4:
A popular
form of
integrated
circuit is the
dual in-line
pin (DIP)
package.

Some ICs are very sensitive to static electricity (which we discuss in Chapter 1), so when you store your ICs, be sure to enclose them in special conductive foam (sold by most electronics suppliers). And before handling an IC, make sure you discharge yourself by touching a conductive material that is connected to earth ground (such as the grounded metal case of your home computer, for example), so you don't zap your IC and wonder why it's not working. (Don't count on the metal pipes in your house to provide a conduit for static charge dissipation. Many home plumbing systems use plastic pipes along the way, so the metal pipes you see in your house aren't necessarily electrically connected to the earth.)

Probing IC pinouts

The pins on an IC package provide connections to the tiny integrated circuits inside, but alas, the pins are not labeled on the package so you have to rely on the datasheet for the particular IC in order to make the proper connections. Among other things, the datasheet provides you with the IC's *pinout,* which describes the function of each pin.

You can find datasheets for most common (and many uncommon) ICs on the Internet. Try using a search engine, such as Google or Yahoo!, to help you locate them.

To determine which pin is which, you look down on the top of the IC (not up at the little critter's underbelly), and look for the *clocking mark* — usually a small notch in the packaging, but it can also be a little dimple, or a white or colored stripe. By convention, the pins on an IC are numbered counterclockwise, starting with the upper-left pin closest to the clocking mark. So, for example, with the clocking notch orienting the chip at the 12 o'clock position, the pins of a 14-pin IC are numbered 1 through 7 down the left side and 8 through 14 up the right side, as shown in Figure 7-5.

Clocking mark

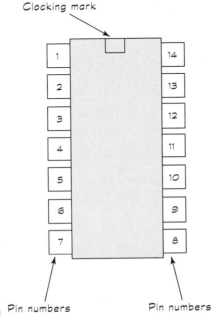

Figure 7-5:
IC pin num-
bering runs
counter-
clockwise
from the
upper left.

Pin numbers   Pin numbers

Don't assume that all ICs with the same number of pins have the same *pin-outs* (arrangement of external connections — in this case, pins), or even that they use the same pins for power connections. And never — *never!* — make random connections to IC pins, under the misguided notion that you can explore different connections until you get the IC to work. That's a sure-fire way to destroy a poor defenseless circuit.

Many circuit diagrams (schematics) indicate the connections to integrated circuits by showing an outline of the IC with numbers beside each pin. The numbers correspond to the counterclockwise sequence of the device's pins, as viewed from the top. (Remember, you start with 1 in the upper left and count up as you go counterclockwise around the chip.) You can easily wire up an IC with these kinds of diagrams because you don't need to look up the device in a book or datasheet. Just make sure that you follow the schematic and that you count the pins properly.

Sourcing and sinking current

Because the insides of integrated circuits are hidden from view, it's hard to know exactly how current flows when you connect a load or other circuitry to the IC's output pin or pins. Typically, datasheets will specify how much current an IC output can *source* or *sink*. An output is said to *source current* when current flows out of the output pin, and *sink current* when current flows into the output pin. If you connect a device, say a resistor, between an output pin and the posi-

tive terminal of a power supply, and the output goes low (0 volts), current will flow through the resistor into the IC — the IC sinks the current. If you connect a resistor between the output pin and the negative supply (ground), and the output goes high, current will flow out of the IC and through the resistor — the IC sources the current. Refer to the datasheet for the maximum source or sink current (which are usually the same value) of an IC output.

If a schematic lacks pin numbers, you need to find a copy of the pinout diagram. For standard ICs, you can find these diagrams in reference books and online; for non-standard ICs, you have to visit the manufacturer's Web site to get the datasheet.

Relying on IC datasheets

IC datasheets are like owner's manuals, providing detailed information about the insides, outsides, and recommended use of an integrated circuit. They are created by the IC manufacturer and are usually several pages long. Typical information contained in a datasheet includes

- Manufacturer's name
- IC name and part number
- Available packaging formats (for instance, 14-pin DIP) and photos of each format
- Dimensions and pinout diagrams
- Brief functional description
- Minimum/maximum ratings (such as power-supply voltages, currents, power, and temperature)
- Recommended operating conditions
- Input/output waveforms (showing how the chip changes an input signal)

Many datasheets include sample circuit diagrams, illustrating how to use the IC in a complete circuit. You can get lots of guidance and good ideas from IC datasheets. Sometimes it really pays to read the owner's manual!

Manufacturers often publish application notes for their integrated circuits. An *application note* (often called an *app note*) is a multi-page document that explains in greater detail than the datasheet how to use the IC in an *application* — a circuit designed for a specific practical task.

Hanging Out with Some Popular ICs

You can find a seemingly endless supply of integrated circuits on the market today, but two in particular are known far and wide for their outstanding versatility and ease of use: the operational amplifier (really a class of ICs) and the 555 timer. It pays to get to know these two circuits fairly well if you intend to get even remotely serious about developing your electronics habit.

In this section, we describe these two popular ICs and one additional IC, the 4017 CMOS decade counter. You encounter the 555 timer IC and the 4017 decade counter IC in projects in Chapter 15, so the upcoming sections provide a quick rundown on how they work.

Operational amplifiers

The most popular type of linear (analog) IC is undoubtedly the *operational amplifier,* nicknamed the *op amp,* which is designed to add muscle to (amplify) a weak signal. An op amp contains several transistors, resistors, and capacitors, and offers more robust performance than a single transistor. For example, an op amp can provide uniform amplification over a much wider range of frequencies *(bandwidth)* than can a single-transistor amplifier.

Most op amps come in 8-pin DIPs (as shown in Figure 7-6), and include two input pins (pin 2, known as the *inverting input,* and pin 3, known as the *non-inverting input*) and one output pin (pin 6). An op amp is one type of *differential amplifier:* The circuitry inside the op amp produces an output signal that is a multiple of the *difference* between the signals applied to the two inputs. Used a certain way, this setup can help eliminate noise (unwanted voltages) in the input signal by subtracting it out of what's amplified.

You can configure an op amp to multiply an input signal by a known gain factor that is determined by external resistors. One such configuration, known as an *inverting amplifier,* is shown in Figure 7-7. The values of the resistors connected to the op amp determine the gain of the inverting amplifier circuit:

$$\text{Gain} = R_2/R_1$$

For instance, if the value of R2 is 10kΩ and that of R1 is 1kΩ, then the gain is 10. With a gain of 10, a 1 V input signal (peak value) produced a 10 V (peak) output signal.

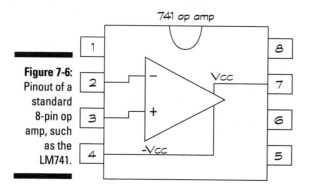

Figure 7-6:
Pinout of a
standard
8-pin op
amp, such
as the
LM741.

To use the inverting amplifier, you just apply a signal (for instance, the
output of a microphone) between the input pins; the signal, amplified several
times, then appears at the output, where it can drive a component (such as a
speaker). Because of the way the op amp in Figure 7-7 is configured, the input
signal is flipped, or *inverted,* to produce the output signal.

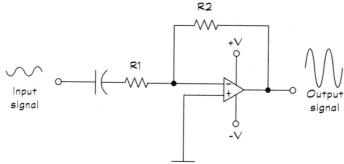

Figure 7-7:
An invert-
ing op-amp
circuit
provides
uniform gain
over a wide
range of
frequencies.

Most op amps require both positive and negative supply voltages. A positive
supply voltage in the range of 8 to 12 V (connected to pin 7) and a negative
supply voltage in the range of –8 to –12 V (connected to pin 4) works. (If you're
looking for some light reading, you can find application notes on how to operate such dual-supply op amps using a single power supply.)

There are gobs of different op amps available at prices ranging from just a
few cents for standard op-amp ICs, such as the LM741 general-purpose op
amp, to more than $100 for high-performance op amps.

IC time machine: The 555 timer

One of the most popular and easy-to-use integrated circuits is the versatile 555 timer, first introduced in 1971 and still in wide use today, with more than one billion units produced every year. This little workhorse can be used for a variety of functions in both analog and digital circuits, most commonly for precision timing (ranging from microseconds to hours), and is the corner-stone of many projects you can build (including several in Chapter 15).

Figure 7-8 illustrates the pin assignments for the 555 timer. Among the pin functions are

- **Trigger input:** When you apply a low voltage to pin 2, you trigger the internal timing circuit to start working. This is known as an *active low* trigger.

- **Output pin:** The output waveform appears on pin 3.

- **Reset:** If you apply a low voltage to pin 4, you reset the timing function, and the output pin (pin 3) goes low. (Some circuits don't use the reset function, and this pin is tied to the positive supply.)

- **Control voltage input:** If you want to override the internal trigger circuit (which normally you don't do), you apply a voltage to pin 5. Otherwise you connect pin 5 to ground, preferably through a 0.01 µF capacitor.

- **Threshold input:** When the voltage applied to pin 6 reaches a certain level (usually two-thirds the positive power-supply voltage), the timing cycle ends. You connect a resistor between pin 6 and the positive supply. The value of this *timing resistor* influences the length of the timing cycle.

- **Discharge pin:** You connect a capacitor to pin 7. This discharge time of this *timing capacitor* influences the length of the timing intervals.

You can find various models of the 555 timer IC. The 556 timer is a dual version of the 555 timer, packaged in a 14-pin DIP. The two timers inside share the same power supply pins.

By connecting a few resistors, capacitors, and switches to the various pins of the 555 timer, you can get this little gem to perform loads of different functions — and it's remarkably easy to do. You can find detailed, easy-to-read information about its various applications on datasheets. We discuss three popular ways to configure a timing circuit using a 555 here.

Astable multivibrator (oscillator)

The 555 can behave as an *astable multivibrator,* which is just a fancy term to describe a sort of electronic metronome. By connecting components to the chip (as shown in Figure 7-9), you configure the 555 to produce a continuous series of voltage pulses that automatically alternate between low (0 volts) and high (the positive supply voltage, V_s), as shown in Figure 7-10. (The term *astable* refers to the fact that this circuit does not settle down into a stable state, but keeps changing on its own between two different states.) This self-triggering circuit is also known as an *oscillator*.

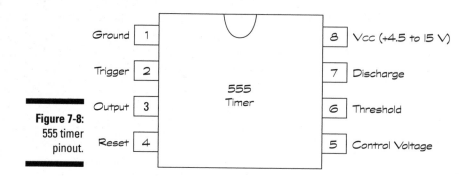

Figure 7-8: 555 timer pinout.

You can use the 555 astable multivibrator for lots of fun things:

- ✔ **Flashing lights:** A low-frequency (< 10 Hz) pulse train can control the on/off operation of an LED or lamp (see the *blinky light* project in Chapter 15).

- ✔ **Electronic metronome:** Use a low-frequency (< 20 Hz) pulse train as the input to a speaker or piezoelectric transducer to generate a periodic clicking sound.

- ✔ **Sounding an alarm:** By setting the frequency to the audio range (20 Hz to 20 kHz) and feeding the output into a speaker or piezoelectric transducer, you can produce a loud, annoying tone (see the *warbler* and *light alarm* projects in Chapter 15).

- ✔ **Clocking a logic chip:** You can adjust the pulse widths to match the specifications for the signal that clocks the logic inside a chip, such as the 4017 decade counter we describe later in this chapter (see the *lighting effects generator* project in Chapter 15).

The frequency f (in hertz), which is the number of complete up-and-down cycles per second, of the square wave produced is determined by your choice of three external components, according to this equation:

$$F = \frac{1.4}{\left(R_1 + 2R_2\right) \times C_1}$$

If you flip the numerator and dominator in that equation, you get the *time period* (T), which is the length of time (in seconds) of one complete up-and-down pulse:

$$T = 0.7 \times (R_1 + 2R_2) \times C_1$$

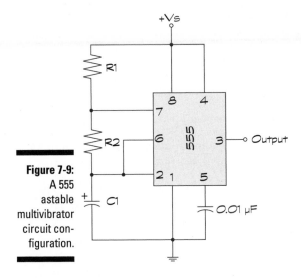

Figure 7-9:
A 555
astable
multivibrator
circuit con-
figuration.

Figure 7-10:
Series of
voltage
pulses
from a 555
astable
multivibra-
tor circuit.
External
components
control the
pulse width.

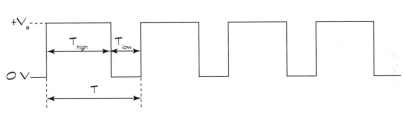

You can set up your circuit so that the width of the high part of the pulse is different from the width of the low part of the pulse. To find the width of the high part of the pulse (expressed as T_{high}), use the following equation:

$$T_{high} = 0.7 \times (R_1 + R_2) \times C_1$$

You find the width of the low part of the pulse (expressed as T_{low}) like this:

$$T_{low} = 0.7 \times R_2 \times C_1$$

If R_2 is much, much bigger than R_1, then the high and low pulse widths will be fairly equal. If $R_2 = R_1$, then the high portion of the pulse will be twice as wide as the low portion. You get the idea.

You can also use a potentiometer (variable resistor) in series with a small resistor as R_1 or R_2 and adjust its resistance to vary the pulses.

To choose values for R_1, R_2, and C_1, we suggest you follow these steps:

1. **Choose C_1.** Decide what frequency range you want to generate and choose an appropriate capacitor. The lower the frequency range, the higher the capacitor you should choose. (Assume that R1 and R2 will be somewhere in the $10k\Omega - 1M\Omega$ range.) For many low-frequency applications, capacitor values of between 0.1 μF and 10 μF work well. For higher-frequency applications, choose a capacitor in the range of 0.01μF to 0.001 μF.

2. **Choose R_2.** Decide how wide the low part of the pulse should be, and choose the value of R_2 that will produce that width, given the value of C_1 you've already determined.

3. **Choose R_1.** Decide how wide the high part of the pulse should be. Using the values of C_1 and R_2 already selected, calculate the value of R_1 that will produce the desired high pulse width.

Monostable multivibrator (one-shot)

By configuring the 555 timer as shown in Figure 7-11, you can use it as a *monostable multivibrator* that generates a single pulse when triggered. It is sometimes called a *one-shot*. Without a trigger, this circuit produces a low (zero) voltage; this is its stable state. When triggered by closing the switch between pin 2 and ground, this circuit generates an output pulse at the level of the supply voltage, V_s. The width of the pulse, T, is determined by the values of R_1 and C_1, as follows:

$$T = 1.1 \times R_1 \times C_1$$

Because capacitor values can often vary by as much as 20 percent, you may need to choose a resistor with a somewhat different value than the formula suggests in order to produce the pulse width you desire.

You can use a one-shot to safely trigger a digital logic device (such as the 4017 CMOS decade counter described later in this chapter). Mechanical switches tend to "bounce" when closed, producing multiple voltage spikes which a digital IC can misinterpret as multiple trigger signals. Instead, if you trigger a one-shot with a mechanical switch, and use the output of the one-shot to trigger the digital IC, you can effectively "de-bounce" the switch.

Bistable multivibrator (flip-flop)

If an astable circuit has no stable state, and a monostable circuit has one stable state, then what's a bistable circuit? If you guessed that a bistable circuit is a circuit with two stable states, you are correct. The 555 *bistable multivibrator* shown in Figure 7-12 produces alternating high (V_s) and low (0 V) voltages, switching from one state to the other only when triggered. Such a circuit is commonly known as a *flip-flop*. There's no need to calculate resistor values; activating the trigger switch controls the timing of the pulses generated.

Figure 7-11: Triggered by closing the switch at pin 2, the 555 monostable circuit produces a single pulse whose width is determined by the values of R_1 and C_1.

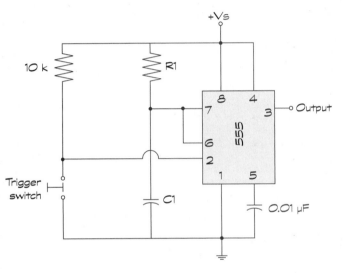

Because it stays low or high until triggered, a flip-flop can be used to store a data bit. (Remember, a bit is a 0 or a 1, which is, respectively, a low or a high voltage.) The registers used to store temporary outputs between stages of logic consist of multiple flip-flops. Flip-flops are also used in certain digital-counter circuits, holding bits in a series of interconnected registers that form an array, the outputs of which make up a bit string representing the count.

You can use various types of 555 timer circuits to trigger other 555 timer circuits. For instance, you can use an oscillator to trigger a flip-flop (useful for clocking registers). Or you can use a one-shot to produce a temporary low-volume tone — and when it ends, change the state of a flip-flop, whose output triggers an oscillator that pulses a speaker on and off. Such a circuit might be used in a home alarm system: Upon entering the home, the homeowner (or intruder) has 10 seconds or so to deactivate the system (while hearing a low-volume warning tone) — or the siren wakes up the neighbors.

Meet the logic families

There are many ways for manufacturers to build digital integrated circuits. A single gate can be constructed using a resistor and a transistor, or just bipolar transistors, or just MOSFETs (another kind of transistor), or other combinations of components. Certain design approaches make it easier to cram lots of tiny gates together in a chip, while other design approaches result in faster circuits or lower power consumption.

Every digital IC is classified according to the design approach and the processing technology used to build its tiny circuits. These classifications are called *logic families*. There are literally dozens of logic families, but the two most famous families are TTL and CMOS.

TTL, or *transistor-transistor logic,* uses bipolar transistors to construct both gates and amplifiers. It is relatively inexpensive to manufacture TTL ICs, but they generally draw a lot of power and require a specific (5-volt) power supply. There are several branches within the TTL family, notably the *Low-Power Schottky* series, which draws roughly one-fifth the power of

conventional TTL technology. Most TTL ICs use the 74*xx* and 74*xxx* format for part numbers, where *xx* or *xxx* specifies a particular type of logic device. For instance, the 7400 is a quad 2-input NAND gate. The Low-Power Schottky version of this part is coded 74LS00.

CMOS, which stands for *complementary metal-oxide semiconductor,* is one type of technology used to make MOSFETs (metal-oxide semiconductor field-effect transistors). (You can see why this family shortened its name to CMOS!) CMOS chips are a little more expensive than their TTL equivalents, but they draw a lot less power and operate over a wider range of supply voltages (3 to 15 volts). They are very sensitive to static electricity, so require special handling. Some CMOS chips are pin-for-pin equivalents for TTL chips, and are identified by a "C" in the middle of the part number. For instance, the 74C00 is a CMOS quad 2-input NAND gate with the same pinout as its cousin, the TTL 7400 IC. Chips in the 40xx series, for instance the 4017 decade counter and 4511 7-segment display driver, are also members of the CMOS family.

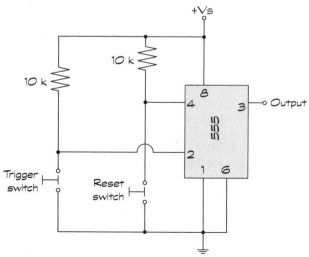

Figure 7-12:
The 555 bistable circuit (or flip-flop) produces a high output when triggered by the switch at pin 2, and a low output when reset by the switch at pin 4.

Counting on the 4017 decade counter

The 4017 CMOS decade counter shown in Figure 7-13 is a 16-pin IC that "counts" from 0 to 9 when triggered. Pins 1–7 and 9–11 go from low to high one at a time when a trigger signal is applied to pin 14. (Of course, they *don't* go from low to high in strict counterclockwise order; you have to check the pinout to determine the order.) You can use the count outputs to light up LEDs (as in the *lighting-effects generator* project in Chapter 15) or trigger a one-shot that controls another circuit.

Counting can take place only when the *disable* pin (pin 13) is low; you can disable counting by applying a high signal to pin 13. You can also force the counter to reset to zero (meaning that the "zero"-count output, which is pin 3, goes high) by applying a high signal (+V) to pin 14.

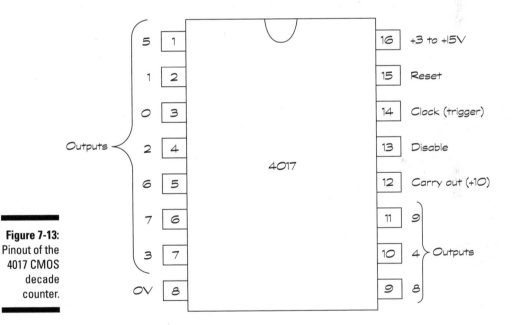

Figure 7-13: Pinout of the 4017 CMOS decade counter.

By piggybacking multiple 4017 ICs together, you can count up tens, hundreds, thousands, and so forth. Pin 12 is high when the count is 0 to 4 and low when the count is 5 to 9, so it looks like a trigger signal that changes at one-tenth the rate of the count. If you feed the output of pin 12 into the trigger input (pin 14) of another decade counter, that second counter will count up tens. By feeding the second counter's pin-12 output into pin 14 of a third counter, you can count up hundreds. With enough 4017 ICs, you may even be able to tally the national debt!

You can also connect two or more of the counter's outputs using diodes to produce a variable timing sequence. To do this, connect each anode (positive side of a diode) to an output pin, and connect all the cathodes (negative sides of the diodes) together and then through a resistor. With this arrangement, when any one of the outputs is high, current will flow through the resistor. For instance, you can simulate the operation of a traffic light by tying outputs 0–4 together and feeding the result (through a resistor) into a red LED, connecting output 5 to a yellow LED, and tying outputs 6–9 together to control a green LED.

Microcontrollers and other popular ICs

Among the other common functions provided by ICs are mathematical operations (addition, subtraction, multiplication, and division), *multiplexing* (selecting a single output from among several inputs), and the conversion of signals between analog and digital:

- ✔ You use an *analog-to-digital (A/D)* converter to convert a real-world analog signal into a digital signal so you can process it with a computer or other digital electronics system.

- ✔ You use a *digital-to-analog (D/A)* converter to convert a processed digital signal back into an analog signal. (For example, you need an analog signal to vibrate the speakers in your home computer system.)

Of course, the *microprocessor* that runs your personal computer (and maybe even your personal life) is also quite popular as ICs go.

Among the most versatile integrated circuits you can find is the microcontroller. A *microcontroller* is a small, complete computer on a chip. To program it, you place it on a development board that allows the IC to interface with your personal computer. After it's programmed, you mount the microcontroller into a socket on your electronic device. You add a few other components — in circuits that provide an interface between the microcontroller and your computer, motors, or switches — and voilà! Your little programmed IC makes things happen (for instance, it can control the motion of a robot). The nice thing about a microcontroller is that you can simply alter a few lines of code (or re-program it completely) to change what it does; you don't need to swap out wires, resistors, and other components in order to get this flexible IC to take on a new personality.

Expanding Your IC Horizons

There's much more to integrated circuits than we can possibly cover in this book. Really smart circuit designers are always coming up with new ideas and improvements on some of the old ideas, so there are a lot of choices in the world of integrated circuits.

You can put your logic to the test by building a simple circuit using a quad 2-input NAND 4011 IC, as shown in Chapter 14. With this learning circuit, you gain experience connecting external circuit components to the pins of an IC chip. Chapter 15 includes several projects involving other ICs.

If you're interested in pursuing the wild IC further, refer to the Appendix for some interesting Web sites that provide useful how-to information for using various ICs in working projects.

Chapter 8

Rounding Out Your Parts List

· ·

· ·

Although the individual components and integrated circuits discussed in Chapters 3 through 7 form the A-team when it comes to shaping the flow of electrons in electronic circuits, there are a bunch of other contributing parts that the A-team relies on to help get the job done.

Some of these other parts — such as wires, connectors, and batteries — are essential ingredients in any electronic circuit. After all, you'd be hard-pressed to build an electronic circuit without wires to connect things together or a source of power to make things run. As for the other parts we discuss in this chapter, you may use them only now and then for certain circuits. For example, when you need to make some noise, a buzzer sure comes in handy — but you may not want to use one in every circuit you build.

In this chapter, we discuss a mixed bag of components, some of which you should keep in stock (just like toilet paper and toothpaste — we hope), while the others can be picked up whenever the spirit moves you.

Making Connections

Making a circuit requires that you connect components to allow electric current to flow between them. The following sections describe wires, cables, and connectors that allow you to do just that.

Choosing wires wisely

Wire that you use in electronics projects is just a long strand of metal, usually made of copper. The wire has only one job: to allow electrons to travel through it. However, you can find a few variations in the types of wire available to you. In the following sections, we give you the lowdown on which type of wire to choose for various situations.

Stranded or solid?

Cut open the cord of any old household lamp (only *after* unplugging the lamp, of course), and you see two or three small bundles of very fine wires, each wrapped in insulation. This is called *stranded* wire. Another type of wire, known as *solid* wire, consists of a single (thicker) wire wrapped in insulation. You can see examples of stranded and solid wires in Figure 8-1.

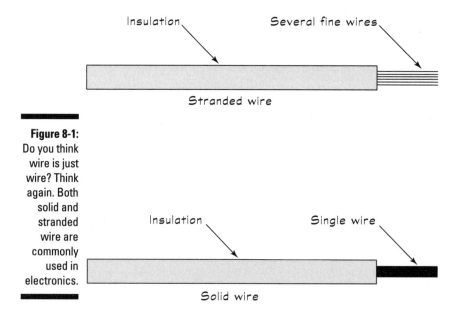

Figure 8-1:
Do you think wire is just wire? Think again. Both solid and stranded wire are commonly used in electronics.

Stranded wire is much more flexible that solid wire, and you use it in situations in which the wire will be moved or bent a lot (such as in line cords for lamps and the cables you hook up to your home entertainment system). You use solid wire in places where you don't plan to move the wire around, and to connect components on breadboards (check out Chapter 11 for more on breadboards). It's easy to insert solid wire into holes in the breadboard, but if you try to use a stranded wire, you have to twist the strands to get all of them into the hole, and you may break a strand or two in the process (Trust us! It happens!), which could short out the circuit.

Sizing up your wire gauge

You refer to the diameter of wire as the *wire gauge*. As luck would have it, the relationship between wire gauge and wire diameter in electronics is essentially backward: The smaller the wire gauge, the larger the wire diameter. You can see common wire gauges in Table 8-1.

Table 8-1	Wires Commonly Used in Electronics Projects	
Wire Gauge	*Wire Diameter (inches)*	*Uses*
16	0.051	heavy-duty electronics applications
18	0.040	heavy-duty electronics applications
20	0.032	most electronics projects
22	0.025	most electronics projects
30	0.01	wire-wrap connections on circuit boards

For most electronics projects, including the ones in Chapters 14 and 15 in this book, you use 20- or 22-gauge wire. If you're hooking up a motor to a power supply, you need to use 16- or 18-gauge wire. As you progress in your electronics dabbling, you may find yourself connecting components on special circuit boards using a technique known as *wire wrapping* (discussed in Chapter 11), which calls for smaller (28- or 30-gauge) wire.

You sometimes see gauge abbreviated in weird and wonderful ways. For example, you may see 20 gauge abbreviated as 20 ga., #20, or 20 AWG (AWG stands for American Wire Gauge).

If you start working on projects involving higher voltage or current than the ones we describe in this book, consult the instructions for your project or an authoritative reference to determine the appropriate wire gauge. For example, the *National Electrical Code* lists the required wire gauges for each type of wiring that you use in a house. Make sure that you also have the right skills and sufficient knowledge of safety procedures to work on such a project.

The colorful world of wires

As with the colorful bands that unlock the secrets of resistor values, the colorful insulation around wire can help you keep track of connections in a circuit. It's common practice, when wiring up a DC circuit (for example, when you work with a breadboard) to use red wire for all connections to positive voltage (+V) and black wire for all connections to negative voltage (–V) or to ground. For AC circuits, use green wire for ground connections. Yellow or orange wire is often used for input signals, such as the signal from a microphone into a circuit. If you keep lots of different colors of wire handy, you can color-code your component connections so it's easier to tell what's going on in a circuit just by glancing at it (unless, of course, you're colorblind).

Collecting wires into cables or cords

Cables are actually groups of two or more wires protected by an outer layer of insulation. Line cords that bring AC power from a wall outlet to an electrical device such as a lamp are cables; so are the cords in the mishmash of connections in your home entertainment system. Cables differ from stranded wires because the wires used in cables are separated by insulation.

Plugging in to connectors

If you look at a cable — say, the one that goes from your computer to your printer — you see that it has metal or plastic doodads on each end. These doodads are called *plugs,* and they represent one kind of *connector.* There are also metal or plastic receptacles on your computer and printer that these cable ends fit into. They represent another kind of connector, called a *socket* or *jack.* The various pins and holes in connectors connect the appropriate wire in the cable to the corresponding wire in the device.

There are many different types of connectors used for various purposes. Among the connectors you're likely to hook up with in your electronics adventures are these:

- A *terminal* and *terminal block* work together as the simplest type of connector. A terminal block contains sets of screws in pairs. You attach the block to the case or chassis of your project. Then, for each wire you want to connect, you solder (or crimp) a wire to a terminal. Next, you connect each terminal to a screw on the block. When you want to connect two wires to each other, simply pick a pair of screws and connect the terminal on each wire to one of those screws.

- *Plugs* and *jacks* that carry audio signals between pieces of equipment, such as a guitar and an amplifier, have cables like the one you see in Figure 8-2. Plugs on each end of the cable connect to jacks on the equipment being connected. These cables contain either one or two signal wires (which carry the actual audio signal) and a metal *shield* surrounding the wires. The metal shield protects the signal wires from electrical interference (known as *noise*) by minimizing the introduction of stray current into the wires.

- You typically use *pin headers* to bring signals to and from circuit boards, which are thin boards designed to house a permanent circuit. Pin headers come in handy for complex electronics projects that involve multiple circuit boards. Most pin headers consist of one or two rows of metal posts attached to a block of plastic which you mount on the circuit board. You connect the pin header to a compatible connector at the end of a *ribbon cable* — a series of insulated wires stuck together side-by-side to form a flat, flexible cable. The rectangular shape of the connector allows easy routing of signals from each wire in the cable to the correct part of the circuit board. You refer to pin headers by the number of pins (posts) they use; for example, you may talk about a 40-pin header.

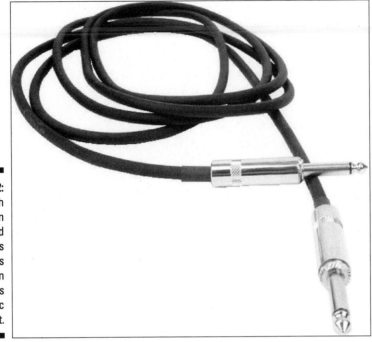

Figure 8-2:
A cable with
a plug on
each end
connects
to jacks
mounted on
two pieces
of electronic
equipment.

Electronics uses a welter of connectors that you don't have to delve into until you start doing more complex projects. If you want to find out more about the broad array of connectors, you can take a look at some of the catalogs or Web sites of electronics suppliers listed in Chapter 17. Most devote an entire category of products to connectors.

Powering Up

All the wires and connectors in the world won't do you much good if you don't have a power source. In Chapter 2, we discuss sources of electricity, including AC power from wall outlets and DC power from batteries and solar cells (also known as photovoltaic cells). Here we discuss how to choose a power source and how to feed their power into your circuits.

Turning on the juice with batteries

For most hobby electronics projects, cells or batteries — which are combinations of cells — are the way to go; the symbol used to represent a cell — and commonly used to represent a battery — in a circuit diagram is shown here. Cells are relatively lightweight and portable, and by combining multiple cells in series, you can create a variety of DC voltage sources. Everyday cells, such as garden-variety AAA-, AA-, C-, and D-cells, all produce about 1.5 volts each. A 9 volt battery (sometimes called a *transistor battery* or PP3 battery) is shaped like a 3-D rectangle and ordinarily contains six 1.5 volt cells. (Some cheap brands may contain only five 1.5 volt cells.) A *lantern battery* (a big boxy thing that can power a flashlight the size of a boom box) produces about 6 volts.

Connecting batteries to circuits

You use a 9 volt (PP3) battery clip (shown in Figure 8-3) to connect an individual 9 volt battery to a circuit. Battery clips snap onto the terminals of the battery (those snaps on the top of the battery are known as a PP3 connector); they contain black and red leads that you connect to your circuit. You strip the insulation off the ends of the black and red wires, and then connect the leads (the bare ends) to your circuit. You can connect the leads to terminals, insert them into holes in a breadboard, or solder them directly to components. We discuss all these techniques in Chapter 11.

When you connect the positive terminal of one battery to the negative terminal of another battery, the total voltage across this series connection is the sum of the individual battery voltages. Battery holders (such as the one shown in Figure 8-4), make series connections between batteries for you while holding multiple batteries in place. Red and black leads from the battery holder provide access to the total voltage. (Some battery holders provide access to the voltage through PP3 connector snaps.)

Figure 8-3:
A battery clip makes it easy to connect a 9 volt battery to your circuit.

Figure 8-4:
Four 1.5 volt batteries in a battery holder produce about 6 volts across the red and black leads.

Sorting batteries by what's inside

Batteries are classified by the chemicals they contain, and the type of chemical determines whether a battery is rechargeable. The following types of batteries are readily available:

✔ **Non-rechargeable batteries:**

- **Zinc-carbon** batteries come in a variety of sizes (AAA, AA, C, D, and 9 volt, among others) and are at the low end of the battery food chain. They may not cost very much, but they also don't last very long.

- **Alkaline** batteries also come in a variety of sizes, and last about three times as long as zinc-carbon batteries. We suggest starting with this type of battery for your projects. If you find yourself replacing them often, you can step up to rechargeable batteries.

- Lightweight **lithium** batteries generate higher voltages — about 3 volts — than other types and have a higher current capacity than alkaline batteries. They cost more, and you can't recharge most lithium batteries, but when your project (for instance, a small robot) calls for a lightweight battery, you can't beat 'em.

✔ **Rechargeable batteries:**

- **Nickel-cadmium (NiCd, pronounced "NYE cad")** batteries generate about 1.2 volts and are the most popular type of rechargeable batteries. Some NiCd batteries still exhibit a flaw known as the *memory effect,* requiring you to fully discharge the battery before recharging it, to ensure that it recharges to its full capacity.

- **Nickel-metal hydride (NiMH)** batteries also generate about 1.2 volts, but don't suffer from the memory effect. We suggest you use NiMH batteries for your projects that need rechargeable batteries.

Be careful not to mix battery types in the same circuit, and *never* attempt to recharge non-rechargeable batteries. These batteries can rupture and leak acid, or even explode. Most non-rechargeable batteries contain warnings about the dangers of such misuse right on their labels.

Buying a recharger and a supply of rechargeable batteries can save you a considerable amount of money over time. Just make sure the battery charger you use is designed for the type of rechargeable battery you select.

Be sure to dispose of batteries properly. Batteries containing heavy metals (such as nickel, cadmium, lead, and mercury) can be hazardous to the environment when improperly disposed of. For guidelines on proper disposal — which may vary from state to state — check battery manufacturer Web sites or other Web sites, such as `http://www.ehso.com/ehshome/batteries.php`.

Rating the life of the everyday battery

The *amp-hour* or *milliamp-hour rating* for a battery gives you an idea of how much current a battery can conduct for a given length of time. For example, a 9 volt battery usually has about a 500 milliamp-hour rating. Such a battery can power a circuit using 25 milliamps for approximately 20 hours before its voltage begins to drop. (We checked a 9 volt battery that we'd used for a few days, and found that it was only producing 7 volts.) An AA battery that has a 1500 milliamp-hour rating can power a circuit drawing 25 milliamps of current for approximately 60 hours.

Six AA batteries in series, which produce about 9 volts, will last longer than a single 9 volt battery. That's because the six series batteries contain more chemicals than the single battery, and can produce more current over time before becoming depleted. (In Chapter 2, we discuss how batteries are made and why they eventually run out of juice.) If you have a project that uses a lot of current, or you plan to run your circuit all the time, consider using larger C- or D-size batteries, which last longer than smaller batteries, or rechargeable batteries.

See the section "Sorting batteries by what's inside," earlier in this chapter, for more about different types of batteries and how long you can expect them to last.

Getting power from the sun

If you're building a circuit designed to operate outside — or you just want to use a clean, green source of energy — you may want to purchase one or more solar panels. A *solar panel* consists of an array of solar cells (which are large diodes known as *photodiodes*) that generate current when exposed to a light source, such as the sun. (We discuss diodes in Chapter 6, and photodiodes in the section "Using Your Sensors," later in this chapter.) A panel measuring about 5 x 5 inches may be able to generate 100 milliamps at 5 volts in bright sunlight. If you need 10 amps, you can certainly get it, but you may find the size of the panel problematic on a small or portable project.

Some solar panels contain output leads that you can connect into your circuit, much like the leads from a battery clip or battery holder. Other solar panels have no leads, so you have to solder your own leads to the two terminals.

Here are some criteria to consider to help you determine whether a solar panel is appropriate for your project:

- ✔ **Do you plan to have the solar panel in sunlight when you want your circuit to be on, or use the panel to charge a storage battery that can power your project?** If not, look for another power source.

- ✔ **Will the solar panel fit on the gadget you're building?** To answer this question, you need to know how much power your gadget will need, and the size of the solar panel that can deliver enough power. If the panel is too large for your gadget, either redesign the gadget to use less power or look for another power source.

Working off your wall power (not recommended)

The AC power supplied by your utility company can cause injury or death if used improperly, so we don't recommend your running circuits directly off household current. And because the vast majority of hobby electronics projects run on batteries, you may never be tempted to work with AC anyway. However, some projects need more current or higher voltages than batteries can easily provide. In those cases, you can use a wall transformer, such as the one shown in Figure 8-5, to convert AC to DC. All the working parts are self-contained in the wall transformer, so you aren't exposed to high AC currents.

Acquiring wall warts

Wall transformers are sometimes called "wall warts" because they stick out of the wall like an ugly wart. You can purchase wall warts new or surplus. (Check out Chapter 17 for some good leads on suppliers.) And, of course, you may already have some old wall warts saved from a discarded cordless phone or other electronic device. If so, be sure to check the voltage and current rating, usually printed on the transformer, to see if it's suitable for your next project. If it is, make sure you know how the connector is wired so you maintain the proper polarity (positive and negative voltage connections) when hooking it up to your circuit.

Wall transformers supply currents ranging from hundreds of milliamps to a few amps at voltages ranging from 5 volts DC to 20 volts DC. Some provide both a positive DC voltage and a negative DC voltage. Different models use different types of connectors to deliver power. If you purchase a wall transformer, be sure to read the specification sheet (specs) carefully to determine how to connect it to your circuit.

Figure 8-5:
A plug-in wall transformer shields you from exposed AC house current.

Switching Electricity On and Off

If you think a switch is just a simple "on/off" mechanism, think again. There are lots of different kinds of switches you can use in electronics projects, and they are categorized by how they are controlled, the type and number of connections they make, and how much voltage and current they can handle.

A *switch* is a device that makes or breaks one or more electrical connections. When a switch is in the *open position,* the electrical connection is broken and you have an open circuit with no current flowing. When a switch is in the *closed position,* an electrical connection is made and current flows.

Controlling the action of a switch

You will hear switches referred to by names that indicate how the switching action is controlled. You can see some of the many different types of switches in Figure 8-6.

- **Slide switch:** You slide a knob back and forth to open and close this switch, which you can find on many flashlights.

- **Toggle switch:** You flip a lever one way close the switch and the other way to open the switch. You may see labels on these switches: "on" for the closed position; "off" for the open position.

- **Rocker switch:** You press one side of the switch down to open the switch, and the other side of the switch down to close the switch. You find rocker switches on many power strips.

- **Leaf switch:** You press a lever or button to temporarily close this type of switch, which is commonly used in doorbells.

- **Pushbutton switch:** You push a button to change the state of the switch, but how it changes depends on the type of pushbutton switch you have:

  - **Push on/push off buttons:** Each press of the button reverses the position of the switch.

  - **Normally open (NO):** This momentary switch is normally open (off), but if you hold the button down, the switch is closed (on). Once you release the button, the switch becomes open again. This is also known as a *push-to-make switch.*

  - **Normally closed (NC):** This momentary switch is normally closed (on), but if you hold the button down, the switch is open (off). When you release the button, the switch becomes closed again. This is also known as a *push-to-break switch.*

✔ **Relay:** A relay is an electrically controlled switch. If you apply a certain voltage to a relay, an electromagnet within pulls the switch lever (known as the *armature*) closed. You may hear talk of closing or opening the *contacts* of a relay's coil. That's just the term used to describe a relay's switch.

Making the right contacts

Switches are also categorized by how many connections they make when you "flip the switch" and exactly how those connections are made. A switch can have one or more *poles,* or sets of input contacts: A *single-pole switch* has one input contact, whereas a *double-pole switch* has two input contacts. A switch can also have one or more conducting positions, or *throws.* With a *single-throw switch,* you either make or break the connection between each input contact and its designated output contact; a *double-throw switch* allows you to alter the connection of each input contact between each of its two designated output contacts.

Figure 8-6:
From top
to bottom:
two toggle
switches,
a rocker
switch,
and a leaf
switch.

Sound confusing? To help clear things up, take a look at the circuit symbols and descriptions of some common switch varieties:

- **Single-pole single-throw (SPST):** This is your basic on/off switch, with one input contact and one output contact. You either make the connection (switch "on") or break the connection (switch "off").

- **Single-pole double-throw (SPDT):** This on/on switch contains one input contact and two output contacts. It is always on; it just switches the input between two choices of outputs. You use an SPDT switch, or _changeover switch,_ when you want to have a circuit turn one device or another on (for example, a green light to let people know they can enter a room, or a red light to tell them to stay out).

- **Double-pole single-throw (DPST):** This dual on/off switch contains two input contacts and two output contacts, and behaves like two separate "make-or-break" SPST switches operating in sync. In the "off" position, both switches are open and no connections are made. In the "on" position, both switches are closed and connections are made between each input contact and its corresponding output contact.

- **Double-pole double-throw (DPDT):** This dual on/on switch contains two input contacts and four output contacts, and behaves like two SPDT (changeover) switches operating in sync. In one position, the two input contacts are connected to one set of output contacts, and in the other position, the two input contacts are connected to the other set of output contacts. Some DPDT switches have a third position, which disconnects (or breaks) all contacts. You can use a DPDT switch as a _reversing switch_ for a motor, connecting the motor to positive voltage to turn one way, negative voltage to turn the other way, and, if there is a third switch position, zero voltage to stop turning.

Using Your Sensors

When you want to trigger the operation of a circuit as a response to something physical happening (such as a change in temperature), you use electronic components known as _sensors._ Sensors take advantage of the fact that various forms of energy — including light, heat, and motion — can be converted into electrical energy. Sensors are a type of _transducer,_ which is an electronic device that converts energy from one form to another. In this section, we describe some of the more common input transducers, or sensors, used in electronic circuits.

Seeing the light

Many electronic components behave differently depending on the light they are exposed to. Manufacturers make certain versions of components to exploit this light sensitivity, enclosing them in clear cases so you can use them as sensors in equipment such as burglar alarms, smoke detectors, automatic dusk-to-dawn lighting, and safety devices that stop your electrically controlled garage door from descending when a cat runs underneath. You can also use them for communications between your remote control, which sends coded instructions via infrared light using a light-emitting diode (or LED, which we discuss in Chapter 6), and your TV or DVD player, which contains a light-sensitive diode or transistor to receive the coded instructions.

Examples of light-sensitive devices used as sensors include the following:

- ✔ **Photoresistors (or photocells)** are light-dependent resistors (LDRs) made from semiconductor material. They typically exhibit a high resistance (about 1MΩ) in darkness and a fairly low resistance (about 100 Ω) in bright light, but you can use a multimeter (as we describe in Chapter 12) to determine the actual resistances exhibited by a specific photoresistor. The typical photoresistor is most sensitive to visible light, especially in the green-yellow spectrum. The symbol for a photoresistor (which can be installed with current running either way in your circuits) is shown here.

- ✔ **Photodiodes** are sort of the opposite of the light-emitting diodes (LEDs) we discuss in Chapter 6. They conduct current or drop voltage only when exposed to sufficient light, usually in the infrared (not visible) range. Like standard diodes, photodiodes contain two leads: The shorter lead is the cathode (negative end) and the longer lead is the anode (positive end).

- ✔ Most **phototransistors** are simply bipolar junction transistors (as we discuss in Chapter 6), encased in a clear package so that light biases the base-emitter junction. These devices usually contain only two leads (whereas standard transistors contain three leads). That's because you don't need access to the base of the transistor in order to bias it — light does that job for you. Phototransistors amplify differences in the light that strikes them, but from the outside, they look just like photodiodes, so you really have to keep track of which is which.

Take a look at Chapter 15 for some projects that involve light-sensitive components.

Capturing sound with microphones

Microphones are input transducers that convert acoustic energy (otherwise known as sound) into electrical energy. Most use a thin membrane, or *diaphragm,* that vibrates in response to air pressure changes from sound. The vibrations of the membrane are translated into an AC electrical signal in various ways, depending on the type of microphone.

- ✔ In a *condenser microphone,* the vibrating membrane plays the role of one plate of a capacitor, so that variations in sound produce corresponding variations in capacitance.(For more about capacitors, see Chapter 4.)

- ✔ In a *dynamic microphone,* the diaphragm is attached to a movable induction coil located inside a permanent magnet. As sound moves the diaphragm, the coil moves inside the magnetic field produced by the magnet, and a current is induced in the coil. (Chapter 5 has the lowdown on this phenomenon, which is known as *electromagnetic induction.*)

- ✔ In a *crystal microphone,* a special *piezoelectric crystal* is used to convert sound into electrical energy, taking advantage of the *piezoelectric effect,* in which certain substances produce a voltage when pressure is applied to them.

- ✔ In a *fiber-optic microphone,* a laser source directs a light beam toward the surface of a tiny reflective diaphragm. As the diaphragm moves, changes in light reflected off the diaphragm are picked up by a detector, which transforms the differences in light into an electrical signal.

Feeling the heat

A *thermistor* is a resistor whose resistance value changes with changes in temperature. The circuit symbol for a thermistor is shown here. Thermistors have two leads and no polarity, so you don't need to worry about which way you insert a thermistor into your circuit.

There are two types of thermistors:

- ✔ **Negative temperature coefficient (NTC) thermistor:** The resistance of an NTC thermistor decreases with a rise in temperature. This is the more common type of thermistor.

- ✔ **Positive temperature coefficient (PTC) thermistor:** The resistance of an PTC thermistor increases with a rise in temperature.

Using a light sensor to detect motion

Have you ever walked up to someone's dark doorway when suddenly the outdoor lights turn on? Or witnessed a garage door stop descending when a child or wheeled object crosses underneath it? If so, you've seen motion detectors at work. Motion-detection devices commonly use light sensors to detect either the *presence* of infrared light emitted from a warm object (such as a person or animal) or the *absence* of infrared light when an object interrupts a beam emitted by another part of the device.

Many homes, schools, and stores use *passive infrared (PIR) motion detectors* to turn on lights or detect intruders. PIR motion detectors contain a sensor (which usually consists of two crystals), a lens, and a small electronic circuit. When infrared light hits a crystal, it generates an electric charge. Because warm bodies (such as those of most humans) emit infrared light at different wavelengths than cooler objects (such as a wall), differences in the output of the PIR sensor can be used to detect the presence of a warm body. The electronic circuit interprets differences in the PIR sensor output to determine whether or not a moving, warm object is nearby.

(Using two crystals in the PIR sensor enables the sensor to differentiate between events that affect both crystals equally and simultaneously — such as changes in room temperature — and events that affect the crystals differently — such as a warm body moving past first one crystal, and then the other.)

Industrial PIR motion detectors use or control 120-volt circuits and are designed to mount on a wall or on top of a floodlight. For hobby projects using a battery pack, you need a compact motion detector that works with about 5 volts. A typical compact motion detector has three leads: ground, positive voltage supply, and the detector output. If you supply +5 volts to the detector, the voltage on the output lead reads about 0 (zero) volts when no motion is detected or about 5 volts when motion is detected. You can find compact motion detectors through online vendors of security systems, but be sure to buy a *motion detector*, rather than just a *PIR sensor*. The lens included in a motion detector helps the device detect the *motion* of an object rather than just the *presence* of an object.

Suppliers' catalogs typically list the resistance of thermistors as measured at 25 degrees Celsius (77° F). Measure the resistance of the thermistor yourself with a multimeter at a few different temperatures (see Chapter 12 for more about using multimeters). These measurements enable you to *calibrate* the thermistor, or get the exact relationship between temperature and resistance. If you're not sure of a thermistor's type, you can figure that out by identifying whether the value increases or decreases with a rise in temperature.

If you're planning to use the thermistor to trigger an action at a particular temperature, be sure to measure the resistance of the thermistor *at that temperature*.

More energizing input transducers

Many other types of input transducers are used in electronic circuits. Here are three common examples:

- **Antennas:** An antenna senses electromagnetic waves, and transforms the energy into an electrical signal. (It also functions as an *output transducer,* converting electrical signals into light waves.)

- **Pressure or position sensors:** These sensors take advantage of the variable-resistance properties of certain materials when they undergo a deformation. Piezoelectric crystals are one such set of materials.

- **Magnetic tape heads:** These devices read magnetic-field fluctuations on audio and video cassette tapes (as well as the computer floppy disks used by the ancients) and convert them into electrical signals.

Transducers are often categorized by the type of energy conversion they perform, for instance, electroacoustic, electromagnetic, photoelectric, and electromechanical transducers. These amazing devices open up tremendous opportunities for electronic circuits to perform countless useful tasks.

Other ways to take your temperature

In the section "Feeling the heat," this chapter discusses the temperature sensors called thermistors — but there are several other types of temperature sensors. Here's a brief summary of their characteristics:

- **Bimetallic strip:** The thermostat in your house probably uses a coiled metal strip, which shrinks as the temperature cools, to trip a switch and turn your furnace on.

- **Semiconductor temperature sensor:** The most common type of temperature sensor, whose output voltage depends on the temperature, contains two transistors (more about those in Chapter 6).

- **Thermocouple:** A thermocouple contains two wires made of different metals (for example, a copper wire and a wire made of a nickel/copper alloy) that are welded or soldered together at one point. These sensors generate a voltage that changes with temperature. The metals it uses determine how the voltage changes with temperature. Thermocouples can measure high temperatures — several hundred degrees or even over a thousand degrees.

- **Infrared temperature sensor:** This sensor measures the infrared light given off by an object. You use it when your sensor must be located at a certain distance from the object you plan to measure; for example, you use this sensor if a corrosive gas surrounds the object. Industrial plants and scientific labs typically use thermocouples and infrared temperature sensors.

Experiencing the Outcome of Electronics

Sensors, or *input transducers,* take one form of energy and convert it into electrical energy, which is fed into the input of an electronic circuit. *Output transducers* do the opposite: They take the electronic signal at the output of a circuit and convert it into another form of energy — for instance, sound, light, or motion (which is mechanical energy).

You may not realize it, but you're probably very familiar with many devices that really are output transducers. Light bulbs, LEDs, motors, speakers, cathode-ray tubes (CRTs), and other electronic visual displays all convert electrical energy into some other form of energy. Without these puppies, you might create, shape, and send electrical signals around through wires and components all day long, and never reap the rich rewards of electronics. It's only when you transform the electrical energy into a form of energy you can experience (and use) personally that you begin to really enjoy the fruits of your labor.

Speaking of speakers

 Speakers convert electrical signals into sound energy; the circuit symbol for a speaker is shown here .Most speakers consist simply of a permanent magnet, an electromagnet (which is a temporary, electrically controlled magnet), and a vibrating cone. Figure 8-7 shows how the components of a speaker are arranged.

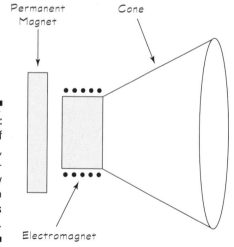

Figure 8-7: The parts of your typical, garden-variety speaker: two magnets and a cone.

The electromagnet, which consists of a coil wrapped around an iron core, is attached to the cone. As electrical current alternates back and forth through the coil, the electromagnet gets pulled toward and then pushed away from the permanent magnet. (Chapter 5 tells you more about the ups and downs of electromagnets.) The motion of the electromagnet causes the cone to vibrate, which creates sound waves.

Most speakers come with two leads that can be used interchangeably. For more serious projects, such as speakers in stereo systems, you must pay attention to the polarity markings on the speakers because of the way they are used in electronic circuits inside the stereo system.

Speakers are rated according to the following criteria:

- ✓ **Frequency range:** Speakers can generate sound over different ranges of frequencies, depending on the size and design of the speakers, within the *audible frequency range* (about 20 Hz [hertz] to 20 kHz [kilohertz]). For example, one speaker in a stereo system may generate sound in the bass range (low audible frequency) while another generates sound in a higher range. You only need to pay close attention to speaker frequency range if you're building a high-end audio system.

- ✓ **Impedance:** Impedance is a measure of the speaker's resistance to AC current (as we discuss in Chapter 5). You can easily find 4Ω, 8Ω, 16Ω, and 32Ω speakers. It's important to select a speaker that matches the minimum impedance rating of the amplifier you're using to drive the speaker. (You can find that rating in the datasheet for the amplifier on your supplier's Web site.) If the speaker impedance is too high, you won't get as much volume out of the speaker as you could, and if the speaker impedance is too low, you may overheat your amplifier.

- ✓ **Power rating:** The power rating tells you how much power (power = current × voltage) the speaker can handle without being damaged. Typical power ratings are 0.25 watt, 0.5 watt, 1 watt, and 2 watt. Be sure you look up the maximum power output of the amplifier driving your speaker (check the datasheet), and choose a speaker with a power rating of at least that value.

For hobby electronics projects, miniature speakers (roughly 2 to 3 inches in diameter) with an input impedance of 8 Ω are often just what you need. Just be careful not to overpower these little noisemakers, which typically handle only $1/4$ to $1/2$ watt.

Sounding off with buzzers

Like speakers, buzzers generate sound — but unlike speakers, buzzers indiscriminately produce the *same* obnoxious sound no matter what voltage you apply (within reason). With speakers, "Mozart in" creates "Mozart out"; with buzzers, "Mozart in" creates nothin' but noise.

 One type of buzzer, a *piezoelectric buzzer,* contains a diaphragm attached to a piezoelectric crystal. The circuit symbol for a piezoelectric buzzer is shown here. When a voltage is applied to the crystal, the crystal expands or contracts (this is known as the piezoelectric effect); this, in turn, makes the diaphragm vibrate, generating sound waves. (Note that this is pretty much exactly the opposite of the way a crystal microphone works, as described earlier in this chapter.)

Buzzers have two leads and come in a variety of packages. Figure 8-8 shows a couple of typical buzzers. To connect the leads the correct way, remember that the red lead connects to a positive DC voltage.

When shopping for a buzzer, you should consider three specifications:

- ✔ **The frequency of sound it emits:** Most buzzers give off sound at one frequency, somewhere in the range of 2 kHz to 4 kHz.

- ✔ **The operating voltage and voltage range:** Make sure you get a buzzer that works with the DC voltage that your project supplies.

- ✔ **The level of sound it produces in unit of decibels (dB):** The higher the decibel rating, the louder (and more obnoxious) the sound emitted. Higher DC voltage provides a higher sound level.

 Be careful that the sound doesn't get so loud that it damages your hearing. You can suffer permanent hearing loss if exposed to sound at 90 dB or higher for a sustained interval — but you won't feel pain until sound reaches at least 125 dB.

Figure 8-8:
These noisy little buzzers are very simple to operate.

Creating good vibrations with DC motors

Have you ever wondered what causes a pager to vibrate? No, it's not Mexican jumping beans: These devices normally use a *DC motor.* DC motors change electrical energy (such as the energy stored in a battery) into motion. That motion may involve turning the wheels of a robot that you build or shaking your pager. In fact, you can use a DC motor in any project in which you need motion.

Electromagnets make up an important part of DC motors because these motors consist of, essentially, an electromagnet on an axle rotating between two permanent magnets, as you can see in Figure 8-9.

The positive and negative terminals of the battery connect so that each end of the electromagnet has the same polarity as the permanent magnet next to it. Like poles of magnets repel each other. This repelling action moves the electromagnet and causes the axle to spin. As the axle spins, the positive and negative connections to the electromagnet swap places, so the magnets continue to push the axle around. A simple mechanism — consisting of a *commutator* (a segmented wheel with each segment connected to a different end of the electromagnet) and brushes that touch the commutator — causes the connections to change. The commutator turns with the axle and the brushes are stationary, with one brush connected to the positive battery terminal and the other brush to the negative battery terminal. As the axle — and (therefore) the commutator — rotates, the segment in contact with each brush changes. This, in turn, changes which end of the electromagnet is connected to negative or positive voltage.

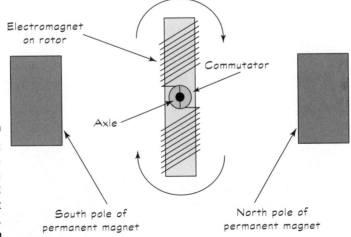

Figure 8-9: How the parts of a simple DC motor fit together.

If you want to get a feel for the mechanism inside a DC motor, buy a cheap one for a few dollars and tear it apart.

The axle in a DC motor rotates a few thousand times per minute — a bit fast for most applications. Suppliers sell DC motors with something called a *gear head* pre-mounted; this device reduces the speed of the output shaft to under a hundred revolutions per minute (rpm). This is similar to the way that changing gears in your car changes the speed of the car.

Suppliers' catalogs typically list several specifications for the motors they carry. When you shop for electric motors, consider these two key characteristics:

✔ **Speed:** The speed (in rpm) that you need depends on your project. For example, when turning the wheels of a model car, you may aim for 60 rpm, with the motor rotating the wheels once per second.

✔ **Operating voltage:** The operating voltage is given as a range. Hobby electronics projects typically use a motor that works in the 4.5 V to 12 V range. Also notice the manufacturer's nominal voltage and stated rpm for the motor. The motor runs at this rpm when you supply the nominal voltage. If you supply less than the nominal voltage, the motor runs slower than the stated rpm. If you supply more, it may run faster but it'll probably burn out.

DC motors have two wires (or terminals that you solder wires to), one each for the positive and negative supply voltage. You run the motor by simply supplying a DC voltage that generates the speed that you want and switching off the voltage when you want the motor to stop.

You can use a more efficient method of controlling the speed of the motor called *pulse-width modulation*. This method turns voltage on and off in quick pulses. The longer the "on" intervals, the faster the motor goes. If you're building a kit for something motor-controlled (such as a robot), this type of speed control should be included with the electronics for the kit.

If you're attaching things such as wheels, fan blades, and so on to the motor shaft, be sure that you have attached the component *securely* before you apply power to the motor. If not, the item may spin off and hit you, or someone near and dear to you, in the face.

Part II
Getting Your Hands Dirty

The 5th Wave   By Rich Tennant

"I swapped out gauges from my treadmill. Want to know how many calories we just burned in the last quarter mile?"

In this part . . .

Ready to get down to business and build an electronic circuit or two (or more)? This part gives you the lowdown on how to set up your own electronics shop, stock up on tools and supplies, and stay safe. It also runs through the essential "how-to's" you need to know so that you can build — and troubleshoot — your circuits.

You find out where (and where not) to locate your electronics workbench, which shiny new tools you absolutely must have, and which electronic components to buy to get started. We explain the language of schematics — circuit diagrams — and how to transfer a circuit design from paper to live, current-carrying components. We show you how to use a solderless breadboard to build prototype (test) circuits and how to solder components to create permanent circuits. You also meet your new best friend — the oh-so-talented multimeter — which enables you to "see" inside your circuits and sniff out problems. And, we explain the basics of two useful, yet optional, test devices: the logic probe and the oscilloscope.

Chapter 9

Setting Up Shop and Ensuring Your Safety

....

In This Chapter

▶ Designing a workspace that works for you

▶ Stockpiling tools and other supplies

▶ Creating a starter kit of electronics components

▶ Realizing that Ohm's Law also applies to the human body

▶ Avoiding electrocution

▶ Keeping your components from turning into lumps of coal

....

*F*inding out about amazing resistors, transistors, and other electronic components is great, but if all you ever do all day is sketch out circuit diagrams and dream about how your cool circuit will manipulate electron flow, you'll never make anything buzz, beep, or move. You need to start tinkering with real components, adding a little power, and watching your circuits perform. But before you go running out to your local electronics supply house, take time to prepare for this circuit-building stage of your life.

In this chapter, we give you guidelines for setting up a little electronics laboratory in your own home. We outline the tools and supplies you need to get the circuit-building job done, and we give you a shopping list of electronic components to purchase so you can build a bunch of different projects.

Because building circuits isn't for the faint of heart (because even small currents can affect your heart), we run you through the safety information you need to know in order to remain a healthy hobbyist. A word to the wise: It doesn't take much electrical current to seriously hurt or even kill you. Even the most seasoned professionals take appropriate precautions to stay safe. We strongly suggest (insist, even) that you thoroughly read the safety information provided (hey, we took great pains putting it together), and before you start each project, review the safety checklist at the end of this chapter.

Promise?

Picking a Place to Practice Electronics

Where you put your workshop is just as important as the projects you make and the tools you use. Just as in real estate, the guiding word for electronics work is location, location, location. By staking out just the right spot in your house or apartment, you'll be better organized and enjoy your experiments much more. There's nothing worse than working with a messy workbench in dim lighting while breathing stale air.

The top ingredients for a great lab

The prime ingredients for the well-set-up electronics laboratory are the following:

- ✔ A comfortable place to work, with a table and chair
- ✔ Good lighting
- ✔ Ample electrical outlets, with at least 15-amp service
- ✔ Tools and toolboxes on nearby shelves or racks
- ✔ A comfortable, dry climate
- ✔ A solid, flat work surface
- ✔ Peace and quiet

The ideal workspace won't be disturbed if you have to leave it for hours or days. Also, the worktable should be off-limits or inaccessible to your children. Curious kids and electronics don't mix!

The garage is an ideal setting because it gives you the freedom to work with solder and other messy materials without worrying about soiling the carpet or nearby furniture. You don't need much space; about 3 by 4 feet ought to do it. If you can't clear that much space in your garage (or you don't have a garage), you can use a room in the house, but try to designate a corner or section of the room for your electronics work. When working in a carpeted room, you can prevent static electricity by spreading a protective cover, such as an anti-static mat, over the floor. We discuss this in detail later in this chapter.

If your work area must be exposed to other family members, find ways to make the area off-limits to others with less knowledge about electronics safety (which we cover later in this chapter), especially young children. Keep your projects, tools, and supplies out of reach or behind lockable doors. And be sure to keep integrated circuits and other sharp parts off the floor — they're painful when stepped on!

No matter where you set up shop, consider the climate. Extremes in heat, cold, or humidity can have a profound effect on your electronics circuits. If you find a work area chilly, warm, or damp, take steps to control the climate in that area, or don't use that area for electronics work. You may need to add insulation, an air conditioner, or a dehumidifier to control the temperature and humidity of your work area. Locate your workbench away from open doors and windows that can allow moisture and extreme temperatures in. And for safety reasons, never — repeat, *never* — work in an area where the floor is wet or even slightly damp.

Workbench basics

The types of projects that you do determine the size of the workbench you need, but for most applications, a table or other flat surface spanning about 2 by 3 feet will suffice. You may even have a small desk, table, or drafting table that you can use for your electronics bench.

You can make your own workbench very easily using an old door as a table surface. If you don't have an old door lying around the house, pick up an inexpensive hollow-core door or a sturdier solid-core door at your local home improvement store. Build legs using 30-inch lengths of 2-by-4 lumber and attach the legs using joist hangers. As an alternative, you can use $^3/_4$-inch plywood or particle board to fashion your work surface.

If you prefer, forgo the 2-by-4 legs and make a simple workbench using a door and two sawhorses. This way, you can take your workbench apart and store it in a corner when you're not using it. Use bungee cords to secure the door to the sawhorses, to prevent accidentally flipping the top of your workbench off the sawhorses.

Remember, as you work on projects, you crouch over your workbench for hours at a time. You can skimp and buy or build an inexpensive worktable, but if you don't already own a good chair, put one on the top of your shopping list. Be sure to adjust the seat for the height of the worktable. A poor-fitting chair can cause backaches and fatigue.

Acquiring Tools and Supplies

Every hobby has its special assortment of tools and supplies, and electronics is no exception. From the lowly screwdriver to the high-speed drill, you enjoy playing with electronics much more if you have the right tools and an assortment of supplies, organized and stored so that you can put your hands on them when you need them without cluttering your work area.

This section tells you exactly what tools and supplies you need to have in order to complete basic-to-intermediate electronics projects.

If you have a permanent place in your house to work on electronics, you can hang some of the hand tools mentioned in this section on the wall or a peg-board. Reserve this special treatment for the tools you use the most. You can stash other small tools and some supplies in a small toolbox, which you can keep on your workbench. A plastic fishing tackle box with lots of small compartments and one large section can help you keep your things organized.

Stockpiling soldering equipment

Soldering (pronounced *SOD-er-ing*) is the method you use to make semi-permanent connections between components as you build a circuit. Instead of using glue to hold things together, you use small globs of molten metal called *solder* (pronounced *SOD-er*) applied by a device called a *soldering iron*. The metal provides a conductive physical joint, known as a *solder joint*, between the wires and component leads of your circuit.

You'll be glad to know that you need only some pretty simple tools for soldering. You can purchase a basic, no-frills soldering setup for under $10, but the better soldering tools cost a bit more.

At a minimum, you will need the following basic items for soldering:

- ✔ **Soldering iron:** A *soldering iron,* also called a *soldering pencil,* is a wand-like tool that consists of an insulating handle, a heating element, and a polished metal tip. (See Figure 9-1.) Choose a soldering iron that is rated at 25-30 watts, sports a replaceable tip, and has a three-prong plug so that it will be grounded. Some models allow you to use different size tips for different types of projects, and some include variable controls that allow you to change the wattage. (Both are nice but not absolutely necessary.)

- ✔ **Soldering stand:** The stand holds the soldering iron and keeps the (very hot) tip from coming into contact with anything on your work surface. Some soldering irons come with stands. (Usually, these combos are known as *soldering stations.*) The stand should have a weighted base; if not, clamp it to your worktable so it doesn't tip over. A stand is a must-have — unless you want to burn your project, your desk, or yourself!

- ✔ **Solder:** *Solder* is a soft metal that is heated by a soldering iron, and then allowed to cool, forming a conductive joint. Standard solder used for electronics is *60/40 rosin core,* which contains roughly 60 percent tin and 40 percent lead and has a core of rosin flux. (Avoid solder formulated for plumbing, which corrodes electronic parts and circuit boards.) The wax-like *flux* helps to clean the metals you're joining together, and it improves the molten solder's ability to flow around and adhere to the components and wire, ensuring a good solder joint. Solder is sold in spools, and we recommend diameters of 0.031 inch (22 gauge) or 0.062 inch (16 gauge) for hobby electronics projects.

The lead content in 60/40 rosin core solder may pose a health hazard if you don't handle it carefully. Be sure to keep your hands away from your mouth and eyes whenever you've been touching this solder. Above all, don't use your teeth to hold a piece of solder while your hands are busy.

We recommend you also get your hands on these additional soldering tools and accessories:

- ✔ **Wetted sponge:** You use this to wipe off excess solder and flux from the hot tip of the soldering iron. Some soldering stands include a small sponge and a built-in space to hold it, but a clean household sponge also works fine.

- ✔ **Solder removal tools:** A *solder sucker,* also known as a *desoldering pump,* is a spring-loaded vacuum you can use to remove a solder joint or excess solder in your circuit. To use it, melt the solder that you wish to remove, quickly position the solder sucker over the molten blob, and activate it to "suck up" the solder. Alternatively, you can use a *solder* (or *desoldering*) *wick* or *braid,* which is a flat, woven copper wire that you place over unwanted solder and apply heat to. When the solder reaches its melting point, it adheres to the copper wire, which you then remove and dispose of.

- ✔ **Tip cleaner paste:** This gives your soldering tip a good cleaning.

- ✔ **Rosin flux remover:** Available in a bottle or spray can, use this after soldering to clean any remaining flux and prevent it from oxidizing (or rusting, in unscientific terms) your circuit, which can weaken the metal joint.

- ✔ **Extra soldering tips:** For most electronics work, a small ($3/_{64}$-inch through $7/_{64}$-inch radius) conical or chiseled tip, or one simply described as a fine tip, works well, but you can also find larger or smaller tips used for different types of projects. Be sure to purchase the correct tip for your make and model of soldering iron. Replace your tip when it shows signs of corrosion, pitting, or plating that is peeling off; a worn tip doesn't pass as much heat.

In Chapter 11, we explain in detail how to use a soldering iron.

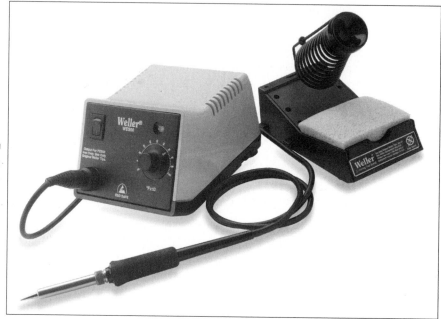

Figure 9-1:
Some soldering iron models are temperature-adjustable and come with their own stands.

Amassing a multimeter

Another essential tool is a *multimeter,* which you use to measure AC and DC voltages, resistance, and current when you want to explore what's going on in a circuit. Most multimeters you find today are of the digital variety (see Figure 9-2), which just means they use numeric displays, like a digital clock or watch. (You can use them to explore analog as well as digital circuits.) An older-style analog multimeter uses a needle to point to a set of graduated scales.

Each multimeter comes with a pair of test leads: one black (for the ground connection) and one red (for the positive connection). On small, pocket units, the test leads are permanently attached to the meters, whereas on larger models, you can unplug the leads. Each test lead has a cone-shaped metal tip used for probing circuits. You can also purchase test clips that slip over the tips, making testing much easier because you can attach these clips onto wires or component leads.

Prices for new multimeters range from $10 to over $100. Higher-priced meters include additional features, such as built-in testing capabilities for capacitors, diodes, and transistors. Think of a multimeter as a set of eyes into your circuits, and consider purchasing the best model you can afford. That way, as your projects grow more complex, you still get a magnificent view of what's going on inside.

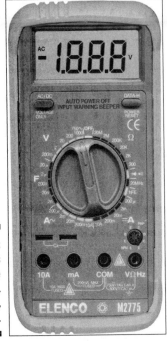

Figure 9-2:
Multimeters
measure
voltages,
resistance,
and current.

If you *really* want an analog multimeter, you may be able to get a top-of-the-line Simpson Model 260 on eBay at a fraction of the (high) price of a new one. The Simpson Model 260 is one of the most popular meters ever produced. It may look like a relic by today's standards, but as long as no one has abused it, it should do all the basic tasks that you need.

We give you the lowdown on how to use a multimeter in Chapter 12.

Hoarding hand tools

Hand tools are the mainstay of any toolbox. These tools tighten screws, snip off wires, bend little pieces of metal, and do all those other mundane tasks. Make sure your have the following tools available at your workbench:

- ✔ **Wire cutter:** You can find general-purpose wire cutters at hardware and home improvement stores, but it pays to invest $5 or so in a flush or *nippy cutter,* shown in Figure 9-3, for making cuts in tight places, such as above a solder joint.

✔ **Wire stripper:** You often need to expose a half-inch or so of bare wire so you can solder a connection or insert the wire into the holes of a solderless breadboard (which we discuss in the next section of this chapter). A good wire stripper contains notches allowing you to neatly and easily strip just the plastic insulation from wires of various sizes (known as gauges, as we describe in Chapter 8), without nicking the copper wire inside. You can also find a combination wire cutter and stripper, but you have to perform your own gauge control.

✔ **Needlenose pliers (two sets):** These pliers help you bend wires, insert leads into breadboard holes, and hold parts in place. Get two pairs: a mini (5-inch long) set for intricate work and a standard-size set to use when you need to apply a wee bit more pressure.

✔ **Precision screwdrivers:** Make sure you have both straight and *Phillips head* (cross-shaped tip) screwdrivers that are small enough for your electronics needs. Use the right size for the job to avoid damaging the head of the screw. A magnetized screwdriver can make it easier to work with small screws, or place a small amount of rubber holdup putty into the head of the screw before inserting the screwdriver tip. Works wonders.

✔ **Magnifying glass:** A 3X (or more) magnifying glass can help you check solder joints and read teeny tiny part numbers.

✔ **Third hand:** No, this isn't a body part from your buddy. It's a tool that clamps onto your worktable and has adjustable clips that hold small parts (or a magnifying glass) while you're working. This makes tasks like soldering a heck of a lot easier. See Figure 9-4 for an example of how to use it.

Collecting cloths and cleansers

If you don't keep the circuitry, components, and other parts of your electronics projects as clean as a whistle, they may not operate as advertised. It's especially important to start with a clean slate if you're soldering parts together or to a circuit board. Dirt makes for bad solder joints, and bad solder joints make for faulty circuits.

Here's a list of items that can help you keep your projects spic-and-span:

✔ **Soft cloth or gauze bandage:** Keep your stuff dust-free by using a soft cloth or sterilized lint-free bandage. Don't use household dusting sprays because some generate static charges that can damage electronics.

✔ **Compressed air:** A shot of compressed air, available in cans, can remove dust from delicate electronic innards. But keep it locked away when you're not using it; if misused as an inhalant, compressed air can cause death.

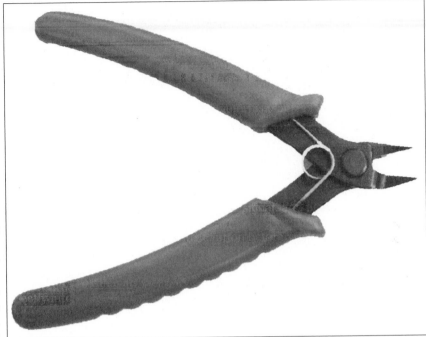

Figure 9-3:
Nippy
cutters trim
wire ends
flush to the
surface.

Figure 9-4:
These
helping
hands
combine
alligator
clips with a
magnifying
glass.

- **Water-based household cleaner:** Lightly spray to remove stubborn dirt and excess grease from tools, work surfaces, and the exterior surfaces of your projects. Don't use them around powered circuits, or you may short something out.

- **Electronics cleaner/degreaser:** Use only a cleaner/degreaser specifically made for use on electronic components.

- **Artist brushes:** Get both a small brush and a wide brush to dust away dirt, but avoid cheap brushes that shed bristles. A dry, clean toothbrush works well, too.

- **Photographic bulb brush:** Available at any photo shop, this combines the whisking action of a soft brush with the cleaning action of a strong puff of air.

- **Contact cleaner:** Available in a spray can, contact cleaner enables you to clean electrical contacts. Spray it onto a brush, and then whisk the brush against the contacts to give them a good cleaning.

- **Cotton swabs:** Soak up excess oil, lubricant, and cleaner with these swabs.

- **Cuticle sticks and nail files:** Scrape junk off circuit boards and electrical contacts, and then give yourself a manicure!

- **Pink pencil eraser:** Great for rubbing electrical contacts clean, especially contacts that have been contaminated by the acid from a leaky battery. Must be pink; other erasers can leave a hard-to-remove residue. Avoid rubbing the eraser against a circuit board because it may create static electricity.

Loading up on lubricants

Motors and other mechanical parts used in electronics projects require a certain amount of grease or oil to operate, and you need to re-lubricate them periodically. There are two types of lubricants commonly used in electronics projects — and one type of lubricant that you should avoid using with electronics projects.

Avoid using a spray-on synthetic lubricant (such as WD-40 and LPS) with your electronics projects. Because you can't control the width of the spray, you're bound to get some on parts that shouldn't be lubed. Also, some synthetic lubricants are non-conductive, and their fine mist can get in the way, interrupting electrical contacts.

The okey-dokey lubricants are:

- ✔ **Light machine oil:** Use this type of oil for parts that spin. Avoid using oil with anti-rust ingredients that may react with plastic parts, causing them to melt. A syringe oiler with a long, thin spout is ideal for hard-to-reach places.

- ✔ **Synthetic grease:** Use lithium grease or another synthetic grease for parts that mesh or slide.

You can find light machine oil and synthetic grease at electronics supply houses as well as many music, sewing machine, hobby, and hardware stores.

Don't apply a lubricant unless you know for sure that a mechanical part needs it. Certain self-lubricating plastics used for mechanical components can break down when exposed to petroleum-based lubricants. If you're fixing a CD player or other piece of electronic equipment, check with the manufacturer for instructions regarding use of lubrication.

Stocking up on sticky stuff

Many electronics projects require that you use an adhesive of some type. For example, you may need to secure a small printed circuit board to the inside of a pocket-sized project box. Depending on the application, you can use one or more of the following adhesives:

- ✔ **White household glue** is best used for projects that involve wood or other porous materials. Allow 20–30 minutes for the glue to dry, and about 12 hours to cure.

- ✔ **Epoxy cement** creates strong, moisture-resistant bonds and can be used for any material. Allow 5–30 minutes for the epoxy to set, and 12 hours for it to cure.

- ✔ **Cyanoacrylate (CA) glue, or super glue,** bonds almost anything (including fingers, so use caution), almost instantly. Use ordinary CA glue when bonding smooth and perfectly matching parts; use the heavier-bodied gap-filling CA glue if the parts don't mate 100 percent.

- ✔ **Double-sided foam tape** is a quick way to secure circuit boards to enclosures or to make sure that loosely fitting components remain in place.

- ✔ **A hot-melt glue gun** allows you to glue things with a drying time of only about 30 seconds. The waterproof, gap-sealing glue comes in a stick that you slide into a slot in the gun, which heats the glue to about 250–350°F — hot enough to hurt you, but not hot enough to melt solder.

Other tools and supplies

There are three other items that we highly recommend you acquire before you begin any electronics work:

- ✔ **Safety glasses:** Stylish plastic safety glasses never go out of fashion. They are a must-have to protect your eyes from flying bits of wire, sputtering solder, exploding electronics parts, and many other small objects. If you wear prescription glasses, place safety glasses over them to ensure complete protection all around your eyes.

- ✔ **Anti-static wrist strap:** This inexpensive strap prevents electrostatic discharge from damaging sensitive electronic components. We discuss this later in this chapter.

- ✔ **First-aid kit and guide:** Burns (or worse) can happen when working with electronic circuits. Keeping a first-aid kit at your workbench is a good idea. Make sure you include guidelines for applying first aid.

There will come a time when you will want to enclose an electronics project in some sort of container with wires or knobs sticking out. For instance, say you build a holiday light display with a controllable blink rate. You may want to place the main circuit in a box, cut a hole through the front of the box, and insert a *potentiometer* (variable resistor) through the hole so you (or someone else) can control how fast the lights blink. Or you may want to build a circuit that detects intruders opening your refrigerator. You could disguise the circuit as a breadbox and place it next to the fridge. In any case, you'll need some additional tools and supplies to enclose your project.

Here's a list of supplies and associated tools you may need in order to box-up your project:

- ✔ **Ready-made box:** You can find simple, unfinished wooden boxes at craft stores, or ABS plastic boxes at most electronics suppliers. Or you can make your own box out of plywood or PVC plastic, using contact cement or another adhesive to keep it together.

- ✔ **Wire clips:** Adhesive-backed plastic clips hold wires in place along the insides of your box.

- ✔ **Cable ties:** Use cable ties to attach wires to non-flat surfaces, such as a wooden dowel.

- ✔ **Electric drill:** A drill with a ³/₈-inch *chuck* (the opening in the drill where you insert the drill bit) comes in handy for making holes in your box for knobs and switches. You can also use it to attach wheels or other external parts to your box.

- ✔ **Hand saws:** You can use a hack saw to cut wood or plastic to make your box, and a coping saw to cut broad openings in the box.

Stocking Up on Parts and Components

Okay, so you've got your workbench set up, complete with screwdrivers, pliers, and hand saws, you've donned your anti-static wrist strap and safety glasses (along with your everyday clothes, please!), and you've got your soldering iron plugged in and ready to go. So what's missing? Oh yeah, circuit components!

When you shop for circuit components, you usually don't go out and purchase just the exact parts listed for a particular circuit diagram, or *schematic*. You purchase an assortment of parts so you can build several different projects without having to run out for parts each time you try something new. Think of this like gathering ingredients for cooking and baking. You keep many basic ingredients, such as flour, sugar, oil, rice, and spices, on hand all the time, and you purchase enough other ingredients to enable you to cook the sorts of things you like for a week or two. Well, it's the same thing with stocking up on electronics parts and components.

In this section, we tell you what parts and how many you should keep on hand in order to build some basic electronics projects.

Solderless breadboards

A *solderless breadboard* is similar, in a way, to a LEGO table: It's a surface on which you can build temporary circuits simply by plugging components into holes arranged in rows and columns across the board. You can very easily take one circuit apart and build another completely different circuit on the same surface.

These holes in a solderless breadboard aren't just ordinary holes; they are *contact holes* with copper lines running underneath so that components plugged into two or more holes within a particular row are connected underneath the surface of the breadboard. You plug in your *discrete components* (resistors, capacitors, diodes, and transistors) and *integrated circuits (ICs)* in just the right way, and — *voilà* — you've got a connected circuit without any soldering necessary. When you're tired of the circuit, you can simply remove the parts and build something else using the same breadboard.

Figure 9-5 shows a small solderless breadboard with a battery-powered circuit connected. The breadboard in the figure has sections of rows and columns connected in a certain way underneath the board. We discuss just how the various contact holes are connected in Chapter 11, where we also discuss how to build circuits using breadboards. For now, just know that different sizes of breadboards with different numbers of contact holes are available.

Figure 9-5:
You can
build a
circuit on
a small
solderless
breadboard
in just
minutes.

A typical small breadboard has 400 contact holes, and it's useful for building smaller circuits with no more than two ICs (plus other discrete components). A typical larger breadboard contains 830 contacts, and you can use it to build somewhat more complex circuits. You can also link multiple breadboards together simply by connecting one or more wires between contact holes on one board and contact holes on the other board.

We recommend that you purchase at least two solderless breadboards, and that at least one of them is a larger (830 contacts) breadboard. Also, buy some adhesive-backed Velcro strips to help hold the breadboards in place on your work surface.

You commonly use solderless breadboards to test your circuit design ideas or explore circuits as you're learning how things work (as you can with the learning circuits in Chapter 14). If you created and test a circuit using a breadboard and you want to use it on a long-term basis, you can re-create the circuit on a soldered or printed circuit board (PCB). A PCB is a kind of breadboard, but instead of contact holes, it has ordinary holes with copper pads surrounding each hole and lines of metal connecting the holes within each row. You make connections by soldering component leads to the copper pads, ensuring the components you're connecting are located in the same row. In this book, we focus exclusively on circuit construction using solderless breadboards.

Circuit-building starter kit

You need an assortment of discrete electronic components (those with two or three individual leads), a few ICs, several batteries, and lots of wire to connect things up. Some components, such as resistors and capacitors, come in packages of ten or more individual pieces. You'll be happy to know that all these components are really inexpensive (cheap, even), but when you add it all up, you'll find it'll cost you one or two weeks' worth of lattes to stock up.

You may wish to refresh your memory on what these components are and how they work by referring to other chapters in this book. Resistors and potentiometers are covered in Chapter 3, capacitors in Chapter 4, and diodes (including LEDs) and transistors in Chapter 6. Integrated circuits (ICs) are covered in Chapter 7, and batteries and wires in Chapter 8.

Here are the discrete components we recommend you start with:

- **Fixed resistors ($^1/_4$-watt or $^1/_2$-watt carbon film):** 10–20 (1 or 2 packages) of each of these resistances: 1kΩ, 10kΩ, 100kΩ, 1MΩ, 2.2kΩ, 22kΩ, 220kΩ, 33kΩ, 470Ω, 4.7kΩ, 47kΩ, 470kΩ.

- **Potentiometers:** Two each of 10kΩ, 100kΩ, 1MΩ.

- **Capacitors:** 10 each (1 package) of 0.01 µF and 0.1 µF non-polarized (polyester or ceramic disc); 10 each (1 package) of 1 µF, 10 µF, 100 µF electrolytic; 3–5 each of 220 µF and 470 µF electrolytic.

- **Diodes:** One each of 1N4001 (or any 1N400x) rectifier diode, 1N4148 small-signal diode, 1 4.3-volt Zener diode (or other Zener breakdown voltage between 3 and 7 volts).

- **LEDs (light emitting diodes):** 10 each (1 package) of red, yellow, and green 5mm diffused LEDs.

- **Transistors:** 3–5 general-purpose, low-power bipolar transistors (such as the 2N3904 NPN or the 2N3906 PNP) and 3–5 medium-power bipolar transistors (such as the NTE 123A NPN or NTE159M PNP). (We use the 2N3906 in a project in Chapter 15.)

We suggest you obtain a few of these popular ICs:

- **555 timer IC:** Get 3–5 of these. You'll use 'em!

- **Op-amp ICs:** Get one or two op amps, such as the LM741 general-purpose amplifier.

- **4017 CMOS decade counter IC:** One will do. (We use one in a project in Chapter 15.) Get two if you want to make a tens counter, too, as we discuss in Chapter 7, or if you think you might accidentally zap the first one with electrostatic discharge.

Don't forget these essential power and wire components:

- **Batteries:** Pick up an assortment of 9 V batteries as well as some 1.5 V batteries. (Size depends on how long you think you're going to run your circuit.)

- **Battery clips and holders:** These devices connect to batteries and provide wire leads to make it easy for you to connect battery power to your circuit. Get 3–5 clips for the size batteries you plan to use.

- **Wire:** Ample 20–22 gauge solid wire. You can buy a 100-foot roll in any one of a variety of colors for about $7. You cut it to various lengths and strip the insulation off each end to connect components. You can solder each end to a component lead, or insert each end into contact holes on your solderless breadboard. Some electronics suppliers sell kits containing dozens of pre-cut, pre-stripped *jumper wires* of various lengths and colors, ideal for use in solderless breadboards. A kit with 140–350 jumper wires may cost you $8–12, but it can save you the time (and trouble) of cutting and stripping your own wire. (Plus, you get rainbow colors!)

You can use a jumper wire as a sort of on/off switch in your circuit, connecting or disconnecting power or components. Just place one end of the jumper wire in your solderless breadboard and alternatively place and remove the other end to make or break the connection.

Adding up the extras

There are lots of other parts and components out there that can enrich your circuits. We recommend you get a few of the ones listed here:

- **Alligator clips:** So-named because they look like the jaws of a fierce 'gator, these insulated clips can help you connect test equipment to component leads, and they can double as heat sinks! Get a bunch (10 or so).

- **Speaker:** You gotta build a circuit that makes noise, so purchase one or two miniature 8-ohm speakers.

- **Switches:** If you think you might enclose one or more projects in a box and you'd like a front-panel on/off control, pick up a couple of SPST (single-pole, single-throw) switches, such as an SPST mini rocker switch for about $1 each. For about $2, you can get one with a built-in LED that lights up when the switch is in the On position.

Organizing all your parts

Keeping all these parts and components organized is essential — unless you're the type who enjoys sorting through junk drawers looking for some tiny, yet important, item. An easy way to get it together is to run over to your local big-box discount store and purchase one or more sets of clear-plastic drawer organizers. Then label each drawer for a particular component (or group of components, such as LEDs, 10–99Ω resistors, and so forth). You'll know in a glance where everything is, and you'll be able to see when your stock is getting low.

Protecting You and Your Electronics

You probably know that Benjamin Franklin "discovered" electricity in 1752 by flying a kite during a lightning storm. Actually, Franklin already knew about electricity and was well aware of its potential power — and potential danger. As Franklin carried out his experiment, he was careful to insulate himself from the conductive materials attached to the kite (the key and a metal wire) and to stay dry by taking cover in a barn. Had he not, we might be looking at someone else's face on the $100 bill!

Respect for the power of electricity is necessary when working with electronics. In this section, you take a look at keeping yourself — and your electronic projects — safe. This is the one section that you really should read from start to finish, even if you already have some experience in electronics.

As you read this section, remember that you can describe electrical current as being one of the following:

- ✔ **Direct current (DC):** The electrons flow one way through a wire or circuit.

- ✔ **Alternating current (AC):** The electrons flow one way, and then another, in a continuing cycle.

Refer to Chapter 2 for more about these two types of electrical current.

Understanding that electricity can really hurt

By far, the single most dangerous aspect of working with electronics is the possibility of electrocution. Electrical shock results when the body reacts to an electrical current — this reaction can include an intense contraction of muscles (namely, the heart) and extremely high heat at the point of contact between your skin and the electrical current. The heat leads to burns that can cause death or disfigurement. Even small currents can disrupt your heartbeat.

The degree to which electrical shock can harm you depends on a lot of factors, including your age, your general health, the voltage, and the current. If you're well over 50 or in poor health, you probably won't stand up to injury as well as if you're 14 and as healthy as an Olympic athlete. But no matter how young and healthy you may be, voltage and current can pack a wallop, so it's important that you understand how much they can harm you.

The two most dangerous electrical paths through the human body are hand-to-hand and left hand to either foot. If electrical current passes from one hand to the other, on its way, it passes through the heart. If current passes from the left hand to either foot, it passes through the heart as well as several major organs.

Seeing yourself as a giant resistor

Your body exhibits some resistance to electrical current, mostly due to the poor conductive qualities of dry skin. The amount of resistance can vary tremendously, depending on body chemistry, level of moisture in the skin, the total path across which resistance is measured, and other factors. You'll see figures ranging anywhere from 50,000 ohms to 1,000,000 ohms of resistance for an average human being. (We discuss what resistance is and how it's measured in Chapter 3.)

If your skin is moist (say you have sweaty hands), you're wearing a metal ring, or you're standing in a puddle, you can bet you've lowered your resistance. Industry figures indicate that such activity can result in resistances as low as 100–300 Ω from one hand to the other, or from one hand to one foot. That's not a whole lot of resistance.

To make matters worse, if you're handling high AC voltages (which you shouldn't be), your skin's resistance — wet or dry — won't help you at all. When you're in contact with a metal, your body and the metal form a capacitor: The tissue underneath your skin is one plate, the metal is the other plate, and your skin is the dielectric. (See Chapter 4 for the lowdown on capacitors.) If that metal wire you're holding is carrying an AC current, the capacitor that is your body acts like a short circuit, allowing current to bypass your skin's resistance. Voltage shocks of more than 240 volts will burn right through your skin, leaving deep third-degree burns at the entry points.

Knowing how voltage and current can harm you

You've seen the signs: WARNING! HIGH VOLTAGE. So you might think that voltage is what causes harm to the human body, but it's actually current that inflicts the damage. So why the warning signs? That's because the higher the voltage, the more current can flow for an equal amount of resistance. And because your body is like a giant resistor, you should shy away from high voltages.

So how much current does it take to hurt the average human being? Not much. Table 9-1 summarizes some estimates of just how much — or how little — DC and 60-Hz (hertz) AC current it takes to affect the human body. Remember that a milliamp (mA) is one one-thousandth of an amp (or 0.001 A). Please note that these are *estimates* (no one has performed experiments on real humans to derive these figures), and that each person is affected differently depending on age, body chemistry, health status, and other factors.

As Table 9-1 shows, the average human body is four to six times more sensitive to AC current than to DC current. Whereas a DC current of 15 mA isn't all that dangerous, 15 mA of alternating current has the potential to cause death.

Table 9-1	Effects of Current on Average Human Body	
Effect	*DC current*	*60-Hz AC current*
Slight tingling sensation	0.6–1.0 mA	0.3–0.4 mA
Noticeable sensation	3.5–5.2 mA	0.7–1.1 mA
Pain felt, but muscle control maintained	41–62 mA	6–9 mA
Pain felt, and unable to let go of wires	51–76 mA	10–16 mA
Difficulty breathing (paralysis of chest muscles)	60–90 mA	15–23 mA
Heart fibrillation (within 3 seconds)	500 mA	65–100 mA

So what does all this mean to you as you pursue your electronics hobby? You probably know enough to stay away from high voltages, but what about getting up close and personal with low voltages? Well, even low voltages can be dangerous — depending on your resistance.

Remember that Ohm's Law (which we cover in Chapter 3) states that voltage is equal to current times resistance:

V (voltage) = I (current) × R (resistance)

Let's say your hands are dry and you aren't wearing a metal ring or standing in a puddle, and your hand-to-hand resistance is about 50,000 ohms. (Keep in mind that your resistance under these dry, ringless conditions may actually be lower.) You can calculate an estimate (repeat: *estimate*) of the voltage levels that might hurt you by multiplying your resistance by the different current levels in Table 9-1. For instance, if you don't want to feel even the slightest tingling sensation in your fingers, you need to avoid coming into contact with wires carrying DC voltages of 30 V (that's 0.6 mA × 50,000 Ω). To avoid involuntary muscle contractions (grabbing the wires), you need to keep AC current below 10 mA, so avoid close proximity to 500 volts AC (VAC) or more.

Now, if you're not so careful, and you wear a ring on your finger while tinkering around with electronics, or you step in a little puddle of water created by a dog or small child, you may accidentally lower your resistance to a dangerous level. If your resistance is 5,000 ohms — and it may be even lower — you'll notice a sensation if you handle just 17.5 VDC (because 5,000 Ω × 0.0035 A = 17.5 V), and you'll lose muscle control and have difficulty breathing if you handle 120 VAC line power (because $\frac{120\,\text{V}}{5,000\,\Omega} = 0.024\,\text{A} = 24\,\text{mA}$).

Household electrical systems in the U.S. and Canada operate at about 120 VAC. This significantly high voltage can, and does, kill. You must exercise *extreme caution* if you ever work with 120 VAC line power.

Until you become experienced in the ways of electronics, you're best off avoiding circuits that run directly off household current. Stick with circuits that run off standard-size batteries, or those small plug-in wall transformers. (You can read about these DC sources in Chapter 8.) Unless you do something silly, like licking the terminal of a 9 V battery (and, yes, that will deliver a shock!), you're fairly safe with these voltages and currents.

The main danger of household current is the effect it can have on your heart muscle. It takes only 65–100 mA to send your heart into fibrillation, which means the muscles contract in an uncontrolled, uncoordinated fashion — and the heart isn't pumping blood. At much lower levels (10–16 mA), AC current can cause severe muscle contractions, so what might start out as a loose grip on a high-voltage wire (just to move it a little bit, or something like that) ends up as a powerful, unyielding grip. Trust us: You won't be able to let go. A stronger grip means a lower resistance (you're just making it easier for electrons to travel through your hand and into your body), and a lower resistance means a higher (often fatal) current. (Situations like this really do happen. The body acts like a variable resistor, with its resistance decreasing sharply as the hands tighten around the wire.)

The potential dangers of DC currents are not to be ignored either. Burns are the most common form of injury caused by high DC current. Remember that voltage doesn't have to come from a power plant to be dangerous. It pays to respect even a 9 V transistor battery: If you short its terminals, the battery may overheat and can even explode. Battery explosions often send tiny battery pieces flying out at high velocities, burning skin or injuring eyes.

Working with AC-powered circuits

Although we strongly recommend that you avoid working with circuits that run directly off household current, we realize you can't always do this. Here are some tips designed to help you avoid electrocution if you choose to work with AC power:

✔ **Use a self-contained power supply.** If your project requires an AC power supply (which converts the AC to lower-voltage DC), using a self-contained power supply, such as a plug-in wall transformer, is much safer than using a homemade power supply. A *wall transformer* is a little black box with plug prongs, such as the one you use to charge your cellphone.

✔ **Keep your AC away from your DC.** Physically separating the AC and DC portions of your circuit can help prevent a bad shock if a wire comes loose.

✔ **Keep AC circuits covered.** A little sheet of plastic works wonders.

✔ **Use the proper fuse.** Don't use a fuse with a too-high rating and never bypass the fuse on any device.

✔ **Double- and triple-check your work before applying power.** Ask someone who knows about circuits to inspect your handiwork before you switch the circuit on for the first time. If you decide to test it further, first remove the power by unplugging the power cord from the wall.

✔ **When troubleshooting a live circuit, keep one hand in your pocket at all times.** By using just one hand to manipulate the testing apparatus, you avoid the situation where one hand touches ground and the other a live circuit, allowing AC to flow through your heart.

✔ **Take care when enclosing your project.** Use a metal enclosure only if the enclosure is fully grounded. You need to use a three-prong electrical plug and wire for this. Be sure to firmly attach the green wire (which is always connected to earth ground) to the metal of the enclosure. If you can't guarantee a fully grounded metal enclosure, use a plastic enclosure. The plastic helps insulate you from any loose wires or accidental electrocution. For projects that aren't fully grounded, use only an isolated power supply, such as a wall transformer.

✔ **Secure all wiring inside your project.** Use a strain relief or a cable mount to secure the AC line cord to your project enclosure so you don't expose a live wire. A *strain relief* (available at hardware stores and electronics suppliers) clamps around the wire and prevents you from tugging the wire out of the enclosure.

✔ **Periodically inspect AC circuits.** Look for worn, broken, or loose wires and components, and promptly make any necessary repairs — with the power off!

✔ **Err on the side of caution.** Take a lesson from Mr. Murphy, and assume that if something can go wrong, it will. Keep your work environment free of all liquids, pets, and small children. Post a first-aid chart nearby. Don't work when you're tired or distracted. Be serious and focused while you're working around electricity.

One final word. If you simply must work with AC voltages, *don't do it alone*. Make sure you have a buddy — preferably someone *not* named in your will — nearby who is willing and able to dial 911 when you're lying on the ground unconscious. Seriously.

Maximizing your resistance — and your safety

When working with electronics, it pays to maximize your resistance just in case you come into contact with an exposed wire. Make sure any tools you pick up are insulated, so that you add more resistance between you and any voltages you may encounter.

Take simple precautions to ensure your work area starts out dry and stays dry. For example, don't place a glass of water or cup of coffee too close to your work area; if you accidentally knock it over, you may lower your own resistance or short out circuit components.

Keeping a first-aid chart handy

Even if you're the safest person on earth, it's still a good idea to get one of those emergency first-aid charts that include information about what to do in case of electrical shock. You can find these charts on the Internet; try a search for *first aid wall chart*. You can also find them in school and industrial supply catalogs.

Helping someone who has been electrocuted may require cardio-pulmonary resuscitation (CPR). Be sure that you're properly trained before you administer CPR on anyone. Check out `www.redcross.org` to get more information about CPR training.

Soldering safely

The soldering iron you use to join components in an electronics project operates at temperatures in excess of 700 degrees Fahrenheit. (You can read up on soldering in Chapter 11.) That's about the same temperature as an electric stove burner set at high heat. You can imagine how much that hurts if you touch it.

When using a soldering iron, keep the following safety tips in mind:

- ✔ **Solder only in a well-ventilated area.** Soldering produces mildly caustic and toxic fumes that can irritate your eyes and throat.

- ✔ **Wear safety glasses when soldering.** Solder has been known to sputter.

- ✔ **Always place your soldering iron in a stand designed for the job.** Never place the hot soldering iron directly on a table or workbench. You can easily start a fire or burn your hands that way.

- ✔ **Be sure that the electrical cord doesn't snag on the table or any other object.** Otherwise, the hot soldering iron can get yanked out of its stand and fall to the ground. Or worse, right into your lap!

✔ **Use the appropriate soldering setting.** If your soldering iron has an adjustable temperature control, dial the recommended setting for the kind of solder that you're using. Too much heat can spoil a good circuit.

✔ **Never solder a *live circuit* (a circuit to which you've applied voltage).** You may damage the circuit or the soldering iron — and you may receive a nasty shock.

✔ **Never grab a tumbling soldering iron.** Just let it fall, and buy a new one if the iron is damaged.

✔ **Consider using silver solder.** If you're concerned about health issues — or tend to stick your fingers in your mouth or rub your eyes a lot — you may want to avoid solders that contain lead. Instead, use silver solder specifically designed for use on electronic equipment. (Never use acid-flux solder in electronics; it wrecks your circuits.)

✔ **Unplug your soldering iron when you're done.**

Avoiding static like the plague

One type of everyday electricity that can be dangerous to people and electronic components is static electricity. It's called *static* because it's a form of current that remains trapped in some insulating body, even after you remove the power source. Static electricity hangs around until it dissipates in some way. Most static dissipates slowly over time, but in some cases, it gets released all at once. Lightning is one of the most common forms of static electricity.

If you drag your feet across a carpeted floor, your body takes on a static charge. If you then touch a metal object, such as a doorknob or a metal sink, the static quickly discharges from your body, and you feel a slight shock. This is known as *electrostatic discharge (ESD),* and can run as high as 50,000 V. The resulting current is small — in the μA range — because of the high resistance of the air that the charges arc through as they leave your fingertips, and it doesn't last very long. So static shocks of the doorknob variety generally don't inflict bodily injury — but they can easily destroy sensitive electronic components.

On the other hand, static shocks from certain electronic components can be harmful. The *capacitor,* an electronic component that stores energy in an electric field, is designed to hold a static charge. Most capacitors in electronic circuits store a very minute amount of charge for extremely short periods of time, but some capacitors, such as those used in bulky power supplies, can store near-lethal doses for several minutes — or even hours.

Use caution when working around capacitors that can store a lot of charge so that you don't get an unwanted shock.

Being sensitive to static discharge

The ESD that results from dragging your feet across the carpet or combing your hair on a dry day may several thousand volts — or higher. Although you probably just experience an annoying tickle (and maybe a bad hair day), your electronic components may not be so lucky. Transistors and integrated circuits (ICs) that are made using metal oxide semiconductor (MOS) technology are particularly sensitive to ESD, regardless of the amount of current.

A MOS device contains a thin layer of insulating glass that can easily be zapped away by 50 V of discharge or less. If you, your clothes, and your tools aren't free of static discharge, that MOS field effect transistor (MOSFET) or complementary MOS (CMOS) IC you planned to use will be nothing more than a useless lump. Because bipolar transistors are constructed differently, they are less susceptible to ESD damage. Other components — resistors, capacitors, inductor, transformers, and diodes — don't seem to be bothered by ESD.

We recommend that you develop static-safe work habits for all the components you handle, whether they're overly sensitive or not.

Minimizing static electricity

You can bet that most of the electronic projects you want to build contain at least some components that are susceptible to damage from electrostatic discharge. You can take these steps to prevent exposing your projects to the dangers of ESD:

- **Use an anti-static wrist strap.** Pictured in Figure 9-6, an anti-static wrist strap grounds you and prevents static build-up. It's one of the most effective means of eliminating ESD, and it's inexpensive (less than $10). To use one, roll up your shirt sleeves; remove all rings, watches, bracelets, and other metals; and wrap the strap around your wrist tightly. Then securely attach the clip from the wrist strap to a proper earth ground connection, which can be the bare (unpainted) surface of your computer case — with the computer plugged in — or simply the ground receptacle of a properly installed wall outlet. Be sure to review the instruction sheet that comes with the strap.

- **Wear low-static clothing.** Whenever possible, wear natural fabrics, such as cotton or wool. Avoid polyester and acetate clothing because these fabrics have a tendency to develop a whole lot of static.

- **Use an anti-static mat.** Available in both table-top and floor varieties, an anti-static mat looks like a sponge, but it's really conductive foam. It can reduce or eliminate the build-up of static electricity on your table and your body.

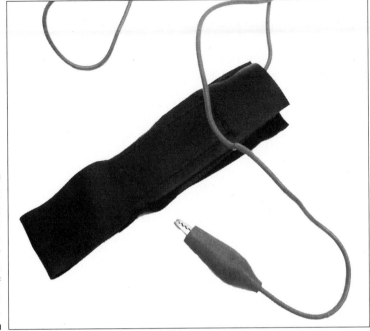

Figure 9-6:
An anti-static wrist strap reduces or eliminates the risk of electrostatic discharge.

Usually, wearing cotton clothing and using an anti-static wrist strap is sufficient for preventing ESD damage.

Grounding your tools

The tools you use when building electronics projects can also build up static electricity — a lot of it. If your soldering iron operates from AC current, ground it to defend against ESD. There's a double benefit here: A grounded soldering iron not only helps prevent damage from ESD but also lessens the chance of a bad shock if you accidentally touch a live wire with the iron.

Cheapo soldering irons use only two-prong plugs and don't have ground connections. Some soldering irons that have three-prong plugs still pose an ESD threat because their tips aren't grounded, even if their bodies are. Because you can't find a really safe and sure means of attaching a grounding wire to a low-end soldering iron, your best bet is to fork over some money ($30 or less) for a new, well-grounded soldering iron. We recommend the Weller WES51 or the Hakko 936, both of which are ESD safe and reasonably affordable. (Hey, you can give up a few lattes to ensure your safety!)

As long as you ground yourself by using an anti-static wrist strap, you generally don't need to ground your other metal tools, such as screwdrivers and wire cutters. Any static generated by these tools is dissipated through your body and into the anti-static wrist strap.

Safety checklist

After you've read all the safety warnings in this chapter, you may wish to review this simple checklist of *minimal* safety requirements before you get started on an electronics project. Better yet, you can make a copy of this checklist, laminate it, and post it at your workbench as a reminder of the simple steps that can ensure your safety — and the well-being of your electronics projects.

Workspace check:

- ✔ Ample ventilation

- ✔ Dry working surface, dry floor

- ✔ No liquids, pets, or small children within a 10-foot range

- ✔ Dangerous tools and materials locked up

- ✔ First-aid chart within view

- ✔ Phone (and caring buddy) nearby

- ✔ Grounded soldering iron with weighted stand

Personal check:

- ✔ Safety glasses

- ✔ Anti-static wrist band (attached to you and to earth ground)

- ✔ No rings, wristwatches, or loose jewelry

- ✔ Cotton or wool clothing

- ✔ Dry hands (or use gloves)

- ✔ Alert and well-rested

Chapter 10

Reading Schematics

. . . .

. . . .

*I*magine driving cross-country without a road map. Chances are, you'd get lost along the way and end up driving in circles. Road maps exist to help you find your way. You can use the equivalent of road maps for building electronic circuits as well. They're called *schematic diagrams,* and they show you how all the parts of the circuits are connected. Schematics show these connections with symbols that represent electronic parts, and lines that show how you attach the parts.

Although not all electronics circuits you encounter are described in the form of a schematic, many are. If you're serious at all about studying electronics, then (sooner or later) you need to understand how to read a schematic. Surprise! The language of schematics isn't all that hard. Most schematic diagrams use only a small handful of symbols for components, such as resistors, capacitors, and transistors.

This chapter tells you all you really need to know to read almost any schematic diagram you come across.

What's a Schematic and Why Should I Care?

A *schematic* is a circuit diagram that shows all the components of a circuit, including power supplies, and their connections. When you're reading a schematic, the most important things to focus on are *the connections,* because the positioning of components in a schematic diagram does not necessarily correspond to the physical layout of components in a constructed circuit. (In fact, for complex circuits, it's highly unlikely that the physical circuit layout reflects the positioning shown in the schematic. Complex circuits often require separate *layout* diagrams, sometimes known as *artwork*.)

Schematics use symbols to represent resistors, transistors, and other circuit components, and lines to show connections between components. By reading the symbols and following the interconnections, you can build the circuit shown in the schematic. Schematics can also help you understand how a circuit operates, which comes in handy when you're testing or repairing the circuit.

Discovering how to read a schematic is a little like learning a foreign language. On the whole, you find that most schematics follow fairly standard conventions. However, just as many languages have different dialects, the language of schematics is far from universal. Schematics can vary depending on the age of the diagram, its country of origin, the whim of the circuit designer, and many other factors.

This book uses conventions commonly accepted in North America. But to help you deal with the variations that you may encounter, we include some other conventions, such as those commonly used in Europe.

Seeing the Big Picture

There's an unwritten rule in electronics about how to orient certain parts of a circuit schematic — especially when drawing diagrams of complex circuits. Batteries and other power supplies are almost always oriented vertically, with the positive terminal on top. In complex schematics, power supplies are split between two symbols (as you will see later), but the positive terminal is usually shown at the top of the schematic (sometimes extending across a horizontal line, or *rail*) and the negative terminal appears at the bottom (sometimes along a rail). Inputs are commonly shown on the left and outputs on the right.

Many electronic systems (for instance, the radio receiver discussed in Chapter 2) are represented in schematics by several stages of circuitry (even though the system really consists of one mighty big circuit). The schematic for such a system shows the sub-circuits for each stage in a left-to-right progression (for instance, the tuner sub-circuit on the left, the detector in the middle, and the amplifier on the right), with the output of the first stage feeding into the input of the second stage, and so forth. Organizing schematics in this way helps make complex circuits more understandable.

It's all about your connections

In all schematics, simple or complex, components are arranged as neatly as possible, and connections within a circuit are drawn as lines, with any bends shown as 90° angles. (No squiggles or arcs allowed!) It's absolutely critical to understand what all the lines in a schematic really mean — and they are not always obvious.

The more complex the schematic, the more likely it is that some lines will crisscross each other (due to the 2-D nature of schematic drawings). You need to know when crossed lines represent an actual wire-it-together connection and when they don't. Ideally, a schematic will clearly distinguish connecting and non-connecting wires like this:

✔ A break or a loop (think of it as a bridge) in one of the two lines at the intersection indicates wires that should *not* be connected.

✔ A dot at the intersection of two lines indicates that the wires *should* be connected.

You can see some common variations in Figure 10-1.

This method of showing connections isn't universal, so you have to figure out which wires connect and which don't by checking the drawing style used in the schematic. If you see an intersection of two lines without a dot to positively identify a real connection, you simply cannot be sure whether the wires should be connected or not. It's best to consult with the person who drew the schematic to determine how to interpret the crisscrossing lines.

To physically implement the connections shown in a schematic, you typically use insulated wires or thin traces of copper on a circuit board. Most schematics don't make a distinction about how you connect the components together; that connection is wholly dependent on how you choose to build the circuit. The schematic's representation of the wiring merely shows you the connections that must be made between components.

Looking at a simple battery circuit

Figure 10-2 shows a simple DC circuit with a 1.5 V battery connected to a resistor labeled R1. The positive side of the battery (+V) is connected to the lead on one side of the resistor; the negative side of the battery is connected to the lead on the other side of the resistor. With these connections made, current flows from the positive terminal of the battery through the resistor, and back to the negative terminal of the battery.

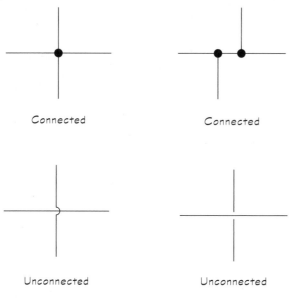

Connected   Connected

Unconnected   Unconnected

Figure 10-1:
You may
encounter
a number of
variations in
how a sche-
matic shows
connections
and non-
connections.

Unknown

Figure 10-2:
A simple
schematic
shows the
connections
between a
battery and
a resistor.

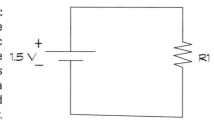

1.5 V

R1

In schematics, "current" is assumed to be *conventional* current, which is described as the flow of positive charges, traveling in a direction opposite to that of real electron flow. (For more about conventional current and electron flow, see Chapter 2.)

Recognizing Symbols of Power

Power for a circuit can come from an alternating current (AC) source, such as the 120 VAC (volts AC) outlet in your house or office (so-called "line power"), or a direct current (DC) source, such as a battery of the low-voltage side of a wall transformer. DC supplies can be positive or negative with respect to the 0-volt reference (known as *common ground*, or simply *common*) in a circuit. Table 10-1 shows various symbols used to represent power and ground connections.

Figuring out the various power connections in a complex schematic is sometimes a task unto itself. This section aims to clear things up a bit. As you read through this section, refer to Table 10-1 to see the symbols as they're discussed.

Table 10-1	Symbols for Power and Ground
Name	*Symbol*
Battery (cell)	
Solar (photovoltaic) cell	
DC power supply	
AC power supply	
Earth ground	
Chassis ground	
Signal ground	

The symbols in Table 10-1 are discussed more fully in the next two subsections.

Showing where the power is

DC power supplies are shown in one of two ways:

✔ **Battery or solar-cell symbol.** Each of the battery symbols in Table 10-1 represents a DC source with two leads. Technically, the battery symbol that includes two parallel lines (the first symbol for "battery") represents a single electrochemical *cell;* the symbol with multiple pairs of lines (the second symbol) represents a *battery* (which consists of multiple cells).

Many schematics use the symbol for a cell to represent a battery.

Each symbol includes a positive terminal (indicated by the larger horizontal line) and a negative terminal. The polarity symbols (+ and –) and nominal voltage are usually shown next to the symbol. The negative terminal is often assumed to be at 0 (zero) volts, unless clearly distinguished as different from the zero-voltage reference (known as *common ground* and detailed later in this chapter). Conventional current flows out of the positive terminal and into the negative terminal when the battery is connected in to a complete circuit.

✔ **"Split" DC power and ground symbols.** To simplify schematics, a DC power supply is often shown using two separate symbols: a small circle at the end of a line representing one side of the supply, with or without a specific voltage label, and the symbol for ground (vertical line with three horizontal lines at the bottom) representing the other side of the supply, with a value of 0 volts. In complex circuits with multiple connections to power, you may see the positive side of the supply represented by a rail labeled +V extending across the top of the schematic. These "split" symbols that represent power supplies are used to eliminate a lot of (otherwise confusing) wire connections in a schematic.

The circuit shown in Figure 10-2 can also be drawn using separate symbols for power and ground, as in Figure 10-3. Note that the circuit in Figure 10-3 is, in fact, a complete circuit.

Many DC circuits use multiple DC power supplies, such as +5 VDC (volts DC), +12 VDC, and even – 5 VDC or – 2 VDC, so the voltage-source symbols in the schematics are usually labeled with the nominal voltage. If a schematic doesn't specify a voltage, you're often (but not always!) dealing with 5 VDC. And remember: Unless otherwise specifically noted, the voltage in a schematic is almost always DC, *not* AC.

Some circuits (for instance, some op-amp circuits, which are discussed in Chapter 7) require both positive and negative DC power supplies. You will often see the positive supply represented by an open circle labeled +V and the negative supply represented by an open circle labeled – V. If the voltages are not specified, they may be +5 VDC and – 5 VDC. Figure 10-4 shows how these power supply connection points are really implemented.

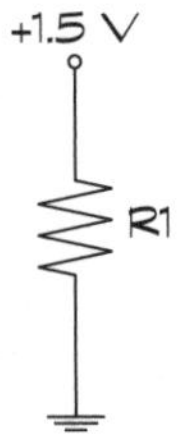

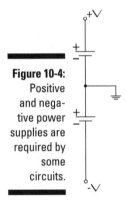

Figure 10-3:
This circuit shows a DC power source symbol at the top and ground symbol at the bottom, which together represent connections to a battery.

Figure 10-4:
Positive and negative power supplies are required by some circuits.

An AC power supply is usually represented by a circle with two leads, either with or without a waveform shape and polarity indicators:

- ✔ **Circle containing waveform:** A squiggly line or other shape inside an open circle represents one cycle of the alternating voltage produced by the power supply. Usually, the source is a sine wave, but it could be a square wave, a triangle wave, or something else.

- ✔ **Circle with polarity:** Some schematics include one or both polarity indicators inside or outside the open circle. This is just for reference purposes, so you can relate the direction of current flow to the direction of voltage swings.

Power for a circuit can come from an AC source, such as the 120-VAC outlet in your house or office (such circuits are called "line-powered"). You typically use an internal power supply to *step down* (or lower) the 120 VAC and convert it to DC. This lower-voltage DC power is then delivered to the components in your circuit. If you're looking at a schematic for a DVD player or some other gadget getting its power from a wall outlet, that schematic probably shows both AC and DC power.

Marking your ground

Ready for some electronics schematic double-talk? When it comes to labeling ground connections in schematics, it's common practice to use the symbol for *earth ground* (which is a real connection to the earth) to represent the *common ground* (the reference point for 0 volts) in a circuit. (Chapter 2 details these two types of grounds.) More often than not, the "ground" points in low-voltage circuits are not actually connected to earth ground; instead, they're just tied to each other — hence the term *common ground* (or simply *common*). Any voltages labeled at specific points in a circuit are assumed to be relative to this common ground. (Remember, voltage is really a differential measurement between two points in a circuit.)

So what symbol should *really* be used for ground points that are not truly connected to the earth? It's the symbol labeled *chassis ground.* Common ground is sometimes called chassis ground because in older equipment, the metal chassis of the device (hi-fi, television, or whatever) served as the common ground connection. Using a metal chassis for a ground connection is not as common today, but we still use the term.

You may also see the symbol for *signal ground* used to represent a zero volt reference point for signals (information-carrying waveforms, discussed in Chapter 2) carried by two wires. One wire is connected to this reference point and the other wire carries a varying voltage representing the signal. Once again, in many schematics, the symbol for earth ground is used instead.

In this book, we use only the schematic symbol for earth ground because most schematics you see these days use that symbol.

As you can see in Figure 10-5, a schematic may show the ground connections in a number of ways:

✔ **No ground symbol:** The schematic can show two power wires connected to the circuit. In a battery-powered circuit, common ground is assumed to be the negative terminal of the battery.

✔ **Single ground symbol:** The schematic shows all the ground connections connected to a single point. It doesn't often show the power source or sources (for instance, the battery), but you should assume that ground connects to the positive or negative DC power sources (as shown in Figure 10-4).

✔ **Multiple ground symbols:** In more complex schematics, it is usually easier to draw the circuit with several ground points. In the actual working circuit, all these ground points connect together.

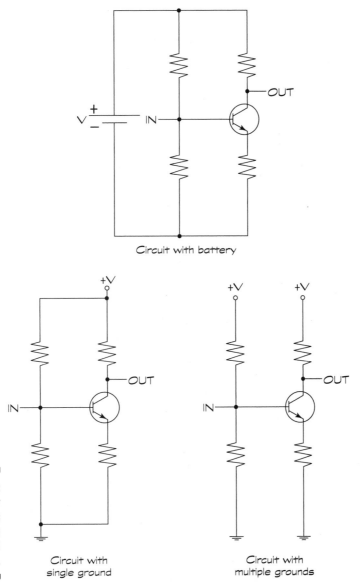

Circuit with battery

Circuit with
single ground

Circuit with
multiple grounds

Labeling Circuit Components

You can find literally hundreds of symbols for electronic components out there, because there are hundreds of component types to depict. Fortunately, you probably encounter only a small number of these symbols in schematics for hobby electronics projects.

Along with the circuit symbol for a particular electronic component, you may see additional information to help uniquely identify the part:

- **Reference ID:** An identifier, such as R1 or Q3. The convention is to use one or more letters to represent the type of component, and a numerical suffix to distinguish one particular component from others of the same type. The most common type designators are R for resistor, C for capacitor, D for diode, L for inductor, T for transformer, Q for transistor, and U or IC for integrated circuit.

- **Part number:** Used if the component is standard (as with a transistor or integrated circuit) or you have a manufacturer's customer product part. For example, a part number may be something like 2N2222 (that's a commonly used transistor) or 555 (a type of IC used in timing applications).

- **Value:** Component values are sometimes shown for passive parts, such as resistors and capacitors, that don't go by conventional part numbers. For example, when indicating a resistor, the value (in ohms) could be marked beside the resistor symbol or the reference ID. Most often, you'll see just the value without a label for the unit of measurement (ohms, microfarads, and so on). Normally resistor values are assumed to be in ohms, and capacitor values are assumed to be in microfarads.

- **Additional information:** A schematic may include additional specifics about one or more components, such as the wattage for a resistor when it isn't your typical $1/4$- or $1/8$- watt value. If you see "10W" next to a resistor value, you know you need a power resistor.

Many schematics show only the reference ID and the circuit symbol for each component, and then include a separate *parts list* to provide the details of part numbers, values, and other information. The parts list maps the reference ID to the specific information about each component.

Analog electronic components

Analog components control the flow of continuous (analog) electrical signals. Table 10-2 shows the circuit symbols used for basic analog electronic components. The third column of the table provides the chapter reference in this book where you can find detailed information about the functionality of each component.

Reference ID primer

Components are often identified in a schematic using an alphabetic type designator, such as C for capacitor, followed by a numerical identifier (1, 2, 3, and so on) to distinguish multiple components of the same type. Together, these identifiers form a *reference ID* that uniquely identifies a specific capacitor or other component. If that value isn't printed beside the component symbol, don't worry; you can find the reference ID in a parts list to indicate the precise value of the component to use. The following type designators are among those most commonly used:

C	Capacitor
D	Diode
IC (or U)	Integrated circuit
L	Inductor
LED	Light-emitting diode
Q	Transistor
R	Resistor
RLY	Relay
T	Transformer
XTAL	Crystal

Table 10-2	Symbols for Analog Components	
Component	*Symbol*	*Chapter Reference*
Resistor		Chapter 3
Variable resistor (potentiometer)		Chapter 3
Photoresistor (photocell)		Chapter 8
Capacitor		Chapter 4
Polarized capacitor		Chapter 4
Variable capacitor		Chapter 4
Inductor		Chapter 5
Air core transformer		Chapter 5

(continued)

Table 10-2 *(continued)*

Component	Symbol	Chapter Reference
Solid core transformer		Chapter 5
Crystal		Chapter 5
NPN (bipolar) transistor		Chapter 6
PNP (bipolar) transistor		Chapter 6
N-channel MOSFET		Chapter 6
P-channel MOSFET		Chapter 6
Phototransistor (NPN)		Chapter 8
Phototransistor (PNP)		Chapter 8
Standard diode		Chapter 6
Zener diode		Chapter 6
Light-emitting diode (LED)		Chapter 6
Photodiode		Chapter 8
Operational amplifier (op amp)		Chapter 7

The circuit symbol for an op amp represents the interconnection of dozens of individual components in a nearly complete circuit (power is external to the op amp). Schematics always use a single symbol to represent the entire circuit, which is packaged as an integrated circuit (IC).The circuit symbol for an op amp is commonly used to represent many other amplifiers, such as the LM386 audio power amplifier used in Chapter 15.

Digital logic and IC components

Digital electronic components — such as logic gates — manipulate digital signals that consist of just two possible voltage levels (high or low). Inside each digital component is a nearly complete circuit (power is external), consisting of individual transistors or other analog components. Circuit symbols for digital components represent the interconnection of several individual components that make up the circuit's logic. You can build the logic from scratch or obtain it in the form of an integrated circuit. Logic ICs usually contain several gates (not necessarily all of the same type) sharing a single power connection.

Table 10-3 shows the circuit symbols for individual digital logic gates. You can find detailed information about the functionality of each logic gate in Chapter 7.

Table 10-3	Symbols for Logic Gates
Component	*Symbol*
AND	
NAND	
OR	
NOR	
XOR	
XNOR	
Inverter (NOT)	

Some schematics show individual logic gates; others show connections to the full integrated circuit, represented by a rectangle. You can see an example of each approach in Figure 10-6. The 74HC00 IC shown in Figure 10-6 is a CMOS quad 2-input NAND gate. In the top circuit diagram, each NAND gate is labeled "$1/4$ 74HC00" because it is one of four NAND gates in the IC. (This type of gate labeling is common in digital circuit schematics.) Note that the fourth NAND gate is not used in this particular circuit (which is why pins 11, 12, and 13 are not used). Whether the schematic uses individual gates or an entire IC package, it usually notes the external power connections. If it doesn't, you have to look up the pinout of the device on the IC datasheet to determine how to connect power. (For more about pinouts and datasheets, see Chapter 7).

You'll find many more digital ICs other than those containing just logic gates. You'll also find linear (analog) ICs that contain analog circuits and mixed signal ICs that contain a combination of analog and digital circuits. Most ICs — except for op amps — are shown the same way in schematics: as a rectangle, labeled with a reference ID (such as IC1) or the part number (such as 74CH00), with numbered pin connections. The function of the IC is usually determined by looking up the part number, but the occasional schematic may include a functional label, such as "one shot."

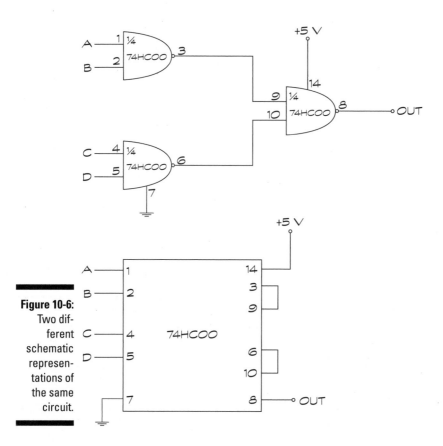

Figure 10-6:
Two different schematic representations of the same circuit.

Miscellaneous components

Table 10-4 lists the symbols for switches and relays. Refer to Chapter 8 for detailed information on each of these components.

Table 10-4	Symbols for Switches and Relays
Component	**Symbol**
SPST switch	
SPDT switch	
DPST switch	
DPDT switch	
Switch, normally open	
Switch, normally closed	
Relay	

Table 10-5 lists the symbols for various input transducers (sensors) and output transducers. (Some of these symbols are cross-referenced in Table 10-2.) You can read about most of these components in Chapter 8, and you can read about LEDs in Chapter 6.

Table 10-5	Symbols for Input and Output Transducers
Microphone	
Thermistor	
Photoresistor (photocell)	

(continued)

Table 10-5 *(continued)*

Photodiode	
Phototransistor (NPN)	
Phototransistor (PNP)	
Antenna	
Incandescent lamp	
Light-emitting diode (LED)	
Speaker	
Piezoelectric buzzer	

Some circuits accept inputs from and send outputs to other circuits or devices. Schematics often show what looks like a loose wire leading into or out of the circuit. Usually it's labeled something like "signal input," or "input from doodad #1," or "output" so you know you're supposed to connect something up to it. (You connect one wire of the signal to this input point, and the other to signal ground.) Other schematics may show a symbol for a specific connector, such as a *plug* and *jack* pair, which connect an output signal from one device to the input of another device. (Chapter 8 provides a closer look.)

Table 10-6 shows a few of the ways input and output connections to other circuits are shown in schematics. Symbols for input/output connections can vary greatly among schematics. The symbols used in this book are among the most common. Although the exact style of the symbol may vary from one schematic to the next, the idea is the same: It's telling you to make a connection to something external to the circuit.

Table 10-6	Symbols for Connections to Other Circuits
Name	**Symbol**
Jack and plug	
Shielded jack	
Unspecified input	
Unspecified output	

Knowing Where to Take Measurements

You may run across a schematic or two that include symbols for test instruments, such as a voltmeter (which measures voltage), an ammeter (which measures current), or an ohmmeter (which measures resistance). (As Chapter 12 explains, a multi-purpose multimeter can function as any of these meters — and more.) You usually see these symbols in schematics on educational Web sites or in documents designed for educational purposes. They point out exactly where to place your meter's leads in order to take the measurement properly.

When you see one of the symbols shown in Table 10-7 in a schematic, remember that it represents a test instrument — not some newfangled "vulcanistor" or other electronic component you've never heard of before.

Table 10-7	Symbols for Common Test Instruments
Name	**Symbol**
Voltmeter	
Ammeter	
Ohmmeter	

Exploring a Schematic

Now that you're familiar with the ABCs of schematics, it's time to put it all together and walk through each part of a simple schematic. The schematic shown in Figure 10-7 shows the LED flasher circuit used in Chapter 15. This circuit controls the on/off blinking of an LED, with the blinking rate controlled by turning the knob of a potentiometer (variable resistor).

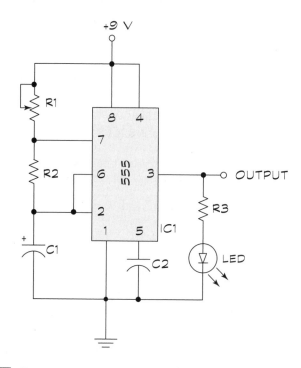

Figure 10-7:
The schematic and parts list used for the blinky-light project in Chapter 15.

Parts List:

IC1: LM555 Timer IC
R1: 1 MΩ potentiometer
R2: 47 kΩ resistor
R3: 330 Ω resistor
C1: 1 µF tantalum capacitor
C2: 9.1 µF disc capacitor
LED: Light-emitting diode

Here's what this schematic is saying:

- At the heart of the schematic is **IC1, an 8-pin 555 timer IC,** with all eight pins connecting to parts of the circuit. Pins 2 and 6 are connected together.

- The circuit is powered by a 9-volt power supply, which can be a **9 V battery.**

  - The positive terminal of the power supply is connected to pins 4 and 8 of IC1, and to one fixed lead and the variable contact (wiper) lead of R1, which is a variable resistor (potentiometer).

  - The negative terminal of the power supply (shown as the common ground connection) is connected to pin 1 of IC1, to the negative side of capacitor C1, to capacitor C2, and to the cathode (negative side) of the LED.

- **R1 is a potentiometer** with one fixed lead connected to pin 7 of IC1 and to resistor R2, and both the other fixed lead and the wiper lead connected to the positive battery terminal (and to pins 4 and 8 of IC1).

- **R2 is a fixed resistor** with one lead connected to pin 7 of IC1 and to one fixed lead of R1, and the other lead connected to pins 2 and 6 of IC1 and the positive side of capacitor C1.

- **C1 is a polarized capacitor.** Its positive side is connected to R2 and to pins 2 and 6 of IC1, and its negative side is connected to the negative battery terminal (as well as to pin 1 of IC1, capacitor C2, and the cathode of the LED).

- **C2 is a non-polarized capacitor** connected on one side to pin 5 of IC1 and on the other side to the negative battery terminal (as well as the negative side of capacitor C1, pin 1 of IC1, and the cathode of the LED).

- The anode of the **LED** is connected to resistor R3 and the cathode of the LED is connected to the negative battery terminal (as well as to the negative side of capacitor C1, capacitor C2, and pin 1 of IC1).

- **R3 is a fixed resistor** connected between pin 3 of IC1 and the anode of the LED.

- Finally, the **output** shown at pin 3 of IC1 can be used as a signal source (input) for another stage of circuitry.

Each item in the walk-through list just given focuses on one circuit component and its connections. Although the list does mention the same connections multiple times, that's consistent with good practice; it pays to check and double-check your circuit connections by making sure *each lead or pin of each individual component* is connected correctly. (Ever hear the rule of thumb "measure twice, cut once?" Well, the same principle applies here.) You can't be too careful when you're connecting electronic components.

Alternative Schematic Drawing Styles

The schematic symbols in this chapter belong to the drawing style used in North America (particularly in the United States) and in Japan. Some countries — notably European nations as well as Australia — use somewhat different schematic symbols. If you're using a schematic for a circuit not designed in the United States or Japan, you need to do a wee bit o' schematic translation in order to understand all the components.

Figure 10-8 shows a sampling of schematic symbols commonly used in the United Kingdom and other European nations. Notice that there are some obvious differences in the resistor symbols, both fixed and variable.

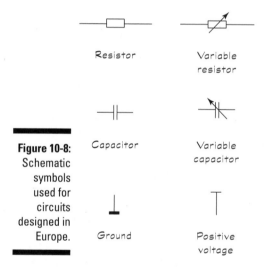

Figure 10-8: Schematic symbols used for circuits designed in Europe.

Resistor   Variable resistor

Capacitor   Variable capacitor

Ground   Positive voltage

This style organizes its symbols differently from the American style. In the United States, you express resistor values over 1,000 Ω in the form of 6.8k or 10.2k, with the lowercase *k* following the value. The European schematic style eliminates the decimal point. Typical of schematics you'd find in the United Kingdom are resistor values expressed in the form 6k8 or 10k2. This style substitutes the lowercase *k* (which stands for *kilohms,* or thousands of ohms) for the decimal point.

You may encounter a few other variations in schematic drawing styles, but all are fairly self-explanatory and the differences are not substantial. After you learn how to use one style of drawing, the others come relatively easily.

Chapter 11

Constructing Circuits

. . . .

In This Chapter

▶ Probing the depths of a solderless breadboard

▶ Creating a no-fuss, no-muss circuit with a solderless breadboard

▶ Soldering — safely — like the pros

▶ Owning up to and fixing soldering mistakes (like the pros)

▶ Solidifying your circuit relationship with a solder breadboard or a perf board

▶ Achieving circuit-building nirvana: your very own custom printed circuit board (PCB)

. . . .

You've carefully set up your workbench, strategically positioned your shiny new toys — oops! we meant *tools* — to impress your friends, shopped around for the best deals on resistors and other components, and now you're ready to get down to business and build some light-flashing, noise-making circuits. So how do you transform an unassuming two-dimensional circuit diagram into a real, live, *working* (maybe even moving) electronic circuit?

In this chapter, we show you various ways to connect electronic components together into circuits that push electrons around at your command. First, we describe how to make flexible, temporary circuits using plug-and-play solderless breadboards, which provide the ideal platform for testing and tweaking your designs. Next, we give you the lowdown on how to safely fuse components together using a molten toxic substance called solder (what fun!). Finally, we outline your options for creating permanent circuits using soldering and/or wire-wrapping techniques, coupled with an assortment of today's most popular circuit boards.

So arm yourself with screwdrivers, needlenose pliers, and a soldering iron, and don your safety glasses and anti-static wristband: You are about to enter the electronics construction zone!

Taking a Look at Solderless Breadboards

Solderless breadboards, also called *prototyping boards* or *circuit breadboards,* make it a snap to build (and dismantle) temporary circuits. These reusable rectangular plastic boards contain several hundred square sockets, or *contact holes,* into which you plug your components (for instance, resistors, capacitors, diodes, transistors, and integrated circuits). Groups of contact holes are electrically connected by flexible metal strips running underneath the surface. You poke a wire or lead into a hole and it makes contact with the underlying metal. By plugging in components in just the right way and running wires from your breadboard to your power supply, you can build a working circuit without permanently bonding components together.

We highly recommend that you use a solderless breadboard (or two) when you first build a circuit. That way, you can test the circuit to make sure it works properly and make any necessary adjustments. Often you can improve on the performance of a circuit just by tweaking a few component values. You can easily make such changes by simply removing one component and inserting another on the board — without having to unsolder and resolder. (For the details of soldering, flip to the "Soldering 101" section later in this chapter.) When you're sure your circuit works the way you want it to, you create a permanent circuit on other types of boards (as described in the section "Making a Commitment: Creating a Permanent Circuit," later in this chapter).

Solderless breadboards are designed for low-voltage DC circuits. Never use a breadboard for 120 VAC house current. Excessive current or voltage can melt the plastic or cause arcing between contacts — ruining the breadboard and possibly exposing you to dangerous currents.

Exploring a solderless breadboard, inside and out

The photo in Figure 11-1 shows a basic solderless breadboard with white lines added to help you visualize the underlying connections between contact holes. In the center of the board, holes are linked vertically in blocks of five (for instance, A, B, C, D, and E in column 30 are all connected together, and F, G, H, I, and J in column 30 are all connected together). There are no connections across the center gap between rows E and F. You can straddle an integrated circuit (IC) across that center gap and instantly set up independent sets of connections for each of its pins.

Across the top and bottom of a breadboard, holes are linked horizontally, but you can't tell exactly how many holes are electrically connected just by looking at the board. For the 400-hole board in Figure 11-1, all 25 contacts in each of the four rows across the top and bottom are electrically connected. In many larger breadboards — such as the 840-hole board pictured in Figure 11-2 — there is a break in the connections halfway across each row. We've placed small jumper wires between neighboring contacts to create a 50-point connection in each row. In some boards, the two rows across the top are electrically connected, as are the two rows across the bottom.

You can use a multimeter to check whether two points within a row — or between rows — are electrically connected. Stick a jumper wire in each hole, and then touch one multimeter probe to one wire and the other probe to the other wire. If you get a low ohm reading, you know that the two points are connected. If you get an infinite ohms reading, you know they aren't connected. (See Chapter 12 for more about testing things with your multimeter.)

Holes are spaced $1/10$ of an inch apart (0.1 inch), a size just right for ICs, most transistors, and discrete components, such as capacitors and resistors. You just plug ICs, resistors, capacitors, transistors, and 20- or 22-gauge solid wire into the proper contact holes to create your circuit. Typically, you use the center two sections of the board to make connections between components; use the top and bottom sections of the board to connect power.

Breadboard manufacturers make contact strips from a springy metal coated with a plating. The plating prevents the contacts from oxidizing, and the springiness of the metal allows you to use wire component leads of different diameters without seriously deforming the contacts. Note, however, that you can damage the contacts if you attempt to use wire larger than 20 gauge or use components with very thick leads. If the wire is too thick to go into the hole, don't try to force it. If you do force it, you can loosen the fit of the contact, and your breadboard may not work the way you want it to.

When you aren't using it, keep your breadboard in a resealable sandwich bag. Why? To keep out the dust. Dirty contacts make for poor electrical connections. Although you can use a spray-on electrical cleaner to remove dust and other contaminants, you make things easier on yourself by keeping the breadboard clean in the first place.

Sizing up solderless breadboard varieties

Solderless breadboards come in many sizes. Smaller breadboards (with 400 to 550 holes) accommodate designs with up to three or four ICs plus a small handful of other discrete components. Larger boards, such as the 840-hole board shown in Figure 11-2, provide more flexibility and accommodate five or more ICs. If you're into really elaborate design work, you can purchase extra-large breadboards with anywhere from 1,660 to more than 3,200 contact holes. These boards can handle one to three dozen ICs plus other discrete components.

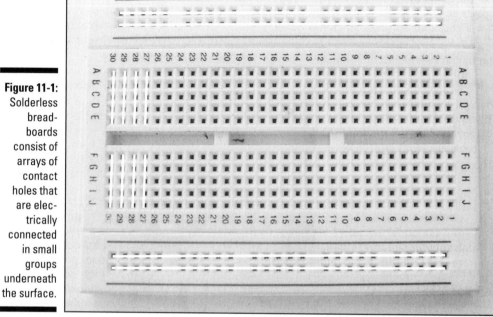

Figure 11-1:
Solderless
bread-
boards
consist of
arrays of
contact
holes that
are elec-
trically
connected
in small
groups
underneath
the surface.

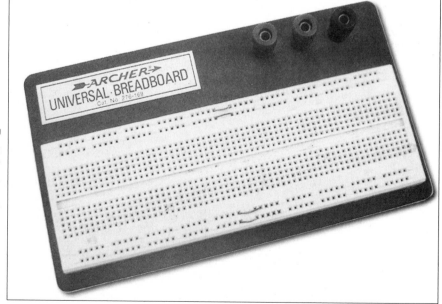

Figure 11-2:
For larger
circuits,
you can
use bigger
solderless
breadboards,
such as this
840-hole
board.

Don't overdo it when buying a solderless breadboard. You don't need a breadboard the size of Wyoming if you're only making small- to medium-sized circuits, such as the ones we show you in Chapter 15. And if you get into the middle of designing a circuit and find that you need a little more breadboard real estate, you can always make connections between two breadboards. Some solderless breadboards even have interlocking ridges so you can put several together to make a larger breadboard.

Building Circuits with Solderless Breadboards

Essentially, breadboarding consists of sticking components into the board, connecting power to the board, and making connections with wire. But there's a right way and a wrong way to do these things. This section gives you the lowdown on what type of wire to use, efficient breadboarding techniques, and how to give your board a neat, logical design.

Preparing your parts and tools

Before you start randomly sticking things into your breadboard, you should make sure you've got everything you need. Check the parts list — the list of electronic ingredients you need to build your circuit — and set aside the components you need. Gather essential tools, such as needlenose pliers, wire cutters, and a wire stripper. Make sure all your component leads are suitable for inserting into breadboard holes. Clip long leads, if possible, so the components will lie flat and snug against the board. (Don't worry if you can't reuse them for another circuit — they're cheap enough.) Some components, such as potentiometers, may not have leads, so you'll need to solder single-core wires to its terminals (see the section "Soldering 101" in this chapter for how to do this). Familiarize yourself with the polarity of parts, which leads are what on transistors, potentiometers, and ICs. And finally, get interconnect wires ready, as described in the very next section.

Saving time with pre-stripped wires

Many of the connections between components on your breadboard are made by the breadboard itself, underneath the surface, but when you just can't make a direct connection via the board, you use interconnect wires (sometimes called *jumper wires*). You use solid (not stranded) 20- to 22-gauge insulated wire to connect components together on your breadboard. Thicker or thinner wire doesn't work well in breadboard: too thick, and the wire won't go into the holes; too thin, and the electrical contact will be poor.

Do not use stranded wire in a breadboard. The individual strands can break off, lodging inside the metal contacts of the breadboard.

While you're buying your breadboard, purchase a set of pre-stripped jumper wires, as we suggest in Chapter 9. (Don't get cheap now; this is worth it.) Pre-stripped wires come in a variety of lengths and are (obviously) already stripped and bent, ready for you to use in breadboards. For instance, one popular assortment contains 10 each of 14 different wire lengths, ranging from $1/10$ inch to 5 inches. A set of 140-350 pre-stripped wires may cost you $8–$12, you can bet the price is well worth the time that you save. The alternative is to buy a bunch of wire, cut segments of various lengths, and painstakingly strip off about $1/3$-inch of the insulation from each end.

Even if you purchase a large assortment of pre-stripped wires, there may come a time when you have to make an interconnect wire or two of your own. You start with 20- or 22-gauge wire (or a longer pre-stripped section that you wish to cut into smaller sections), and cut it to the desired length. If you have a wire stripper with a gauge-selection dial, set the dial for the gauge of wire you're using. Other wire strippers may have several cutting notches labeled for various gauges. Using one of these gauge-specific devices instead of a generic wire stripper prevents you from nicking the wire when you're stripping off the insulation. Nicks weaken the wire, and a weak wire can get stuck inside a breadboard hole and ruin your whole day.

To make your own breadboard wire, follow these steps:

1. **Cut the wire to the length you need, using a wire-cutting tool.**

2. **Strip off about $1/4$- to $1/3$-inch of insulation from each end.**

   If you use a gauge-specific tool, insert one end of the wire into the stripping tool, hold the other end with a pair of needlenose pliers, and draw the wire through the stripping tool. If you use a generic wire stripper, you provide the gauge control by how much you squeeze the tool around the wire: too much, and you nick the wire; too little, and you don't cut through all the insulation.

3. **Bend the exposed ends of wire at a right (90°) angle.** You can use needlenose pliers to do this.

Laying out your circuit

You've got your parts and tools ready, schematic (the circuit diagram) in hand, and now you want to build your circuit on your breadboard. But where should you start? What's the best way to connect everything?

Welcome to the world of circuit layout — figuring out where everything should go on the board so it all fits together and is neat, tidy, and error-free. Don't expect your circuit layout to look exactly like your schematic — that's not only difficult to do, but usually impractical. You can, however, orient your key circuit elements so your circuit is easier to understand and debug.

When you're building a circuit on a breadboard, concentrate on the *connections between components,* rather than on the position of components in your schematic.

Here are some guidelines for building your breadboard circuit:

- ✔ **Use one of the top rails (long rows) for the positive power supply, and one of the bottom rails for ground (and the negative power supply, if there is one).** These rails give you plenty of interconnected sockets so you can easily connect components to power and ground.

- ✔ **Orient any circuit inputs on the left side of the board, and outputs on the right side.** Plan your component layout to minimize the number of jumper wires. The more wiring you have to insert, the more crowded and confusing the board becomes.

- ✔ **Place ICs first, straddling the center gap.** Allow at least three — preferably ten — columns of holes between each IC. You can use a chip inserter/ extractor tool to implant and remove ICs to reduce the chances of damaging the IC while handling it.

  If you're working with CMOS chips, be sure you ground the tool to eliminate stray static electricity.

- ✔ **Work your way around each IC, starting from pin 1, inserting the components that connect to each pin. Then insert any additional components to complete the circuit.** Use needlenose pliers to bend leads and wires to a 90-degree angle and to insert them into the sockets, keeping leads and wires as close to the board as possible to prevent them from getting knocked loose.

- ✔ **If your circuit requires common connection points in addition to power, and you don't have enough points in one column of holes, use longer pieces of wire to bring the connection out to another part of the board where you have more space.** You can make the common connection point one or two columns between a couple of ICs, for instance.

Figure 11-3 shows a resistor, jumper wire, and light-emitting diode (LED) inserted into a breadboard.

Don't worry about urban sprawl on your breadboards. You do better to place components a little farther apart than to jam them too close together. Keeping a lot of distance between ICs and components also helps you to tweak and refine the circuit. You can more easily add parts without disturbing the existing ones.

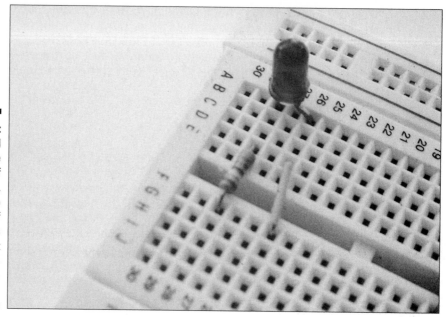

Figure 11-3:
Strip and
bend the
ends of
jump wire,
and clip
the leads of
components
so they fit
neatly on
the bread-
board.

Messy wiring makes it difficult to debug a circuit, and a tangle of wires greatly increases the chance of mistakes. Wires pull out when you don't want them to, or the circuit can malfunction altogether. To avoid chaos, take the time to carefully plan and construct your breadboard circuits. The extra effort can save you lots of time and frustration down the road.

Avoiding damaged circuits

There are just a few other things you need to know to keep your breadboard and circuits in good working order:

✔ **If your circuit uses one or more CMOS chips, insert the CMOS chips last.** If you need to, use a dummy TTL IC to make sure that you wired everything properly. TTL chips aren't nearly as sensitive to static as the CMOS variety. Be sure to provide connections for the positive and nega- tive power supply — and to connect all inputs (tie those inputs that you aren't using to the positive or negative supply rail). When you're ready to test the circuit, remove the dummy chip and replace it with the CMOS IC.

✔ **Never expose a breadboard to heat, because you can permanently damage the plastic.** ICs and other components that become very hot (because of a short circuit or excess current, for example) may melt the plastic underneath them. Touch all the components while you have the circuit under power to check for overheating.

My breadboard circuit doesn't work right!

As you work with solderless breadboards, you may encounter the fairly common problem of *stray capacitance,* which is unwanted capacitance (stored electrical energy) in a circuit. All circuits have an inherent capacitance that can't be avoided, but when there are lots of wires going every which way, the capacitance can unexpectedly increase. At a certain point (and it differs from one circuit to the next), this stray capacitance can cause the circuit to misbehave.

Because solderless breadboards contain strips of metal and require somewhat longer component leads, they tend to introduce a fair amount of stray capacitance into unsuspecting circuits. As a result, solderless breadboards have a tendency to change the characteristics of some components — most notably capacitors and inductors. These variations can change the way a circuit behaves. Keep this fact in mind if you're working with RF (radio-frequency) circuits, such as radio receivers and transmitters, digital circuits that use signals that change at a very fast rate (on the order of a couple of million Hertz), and more sensitive timing circuits that rely on exact component values.

If you're building a radio or other circuit that stray capacitance can affect, you may have to forego the step of first building the circuit on a solderless breadboard. You may have to go straight to a solder breadboard or a perf board (described in the section "Making a Commitment: Creating a Permanent Circuit," later in this chapter).

- ✔ **Never use a solderless breadboard to carry 120 VAC house current.** Current may arc across the contacts, damaging the board and posing a danger.

- ✔ **If a small piece of a lead or wire becomes lodged in a socket, use needlenose pliers to gently pull it out of the hole — with the power switched off.**

- ✔ **You won't always be able to finish and test a circuit in one sitting. If you have to put your breadboard circuit aside for a while, put it out of the reach of children, animals, and the overly curious.**

Soldering 101

Soldering (pronounced "SOD-er-ing") is the method you use to make conductive connections between components and/or wire. You use a device called a *soldering iron* to melt a soft metal called *solder* (pronounced "SOD-er") in a way that the solder flows around the two metal leads you are joining. When the soldering iron is removed, the solder cools and forms a conductive physical joint, known as a *solder joint,* between the wires or component leads.

Should you care about soldering when you plan to use solder*less* breadboards for your circuit construction projects? The answer is yes. Almost every electronics project involves a certain amount of soldering. For instance, you may purchase components (such as potentiometers, switches, and microphones) that don't come with leads — in which case, you have to solder two or more wires to their terminals to create leads so you can connect them to your breadboard.

Of course, you also use this technique extensively when you build permanent circuits on solder boards, perf boards, or printed circuit boards (as described in the section "Making a Commitment: Creating a Permanent Circuit," later in this chapter).

Preparing to solder

To get your soldering jobs done, you need a soldering iron (25–30 watts), a spool of 16- or 22-gauge standard solder (60/40 rosin-core), a secure soldering stand, and a small sponge. (Refer to Chapter 9 for detailed information on how to choose soldering equipment for your electronics projects.) Make sure you secure your soldering iron in its stand and position it in a safe place on your workbench, where it's unlikely to be knocked over.

Gather up a few other items, such as safety glasses (to protect your eyes from sputtering solder), an alligator clip (which doubles as a heat sink for temperature-sensitive components), an anti-static wristband (described in Chapter 9), isopropyl alcohol, a piece of paper, a pencil, and some sticky tape. Place each part that you need to solder on the paper, securing it with sticky tape. Write a label, such as R1, on the paper next to each part so it matches the label on your schematic. Put on your safety glasses and anti-static wristband, and make sure your work area is properly ventilated.

Wet the sponge, squeezing out excess water. Turn on the soldering iron, wait a minute or so for it to heat up (to about 700°F), and then wet the tip of the soldering iron by *briefly* touching it to the sponge. If the tip is new, *tin* it before soldering to help prevent solder from sticking to the tip. (Sticky solder can form an ugly globule, which can wreak havoc if it falls off into your circuit.) You tin the tip by applying a small amount of molten solder to it. Then you wipe off any excess solder on the sponge.

Periodically tin your soldering iron's tip to keep it clean. You can also purchase soldering tip cleaners if dirt becomes caked on and you just can't get it off during regular tip re-tinning.

Soldering for success

Successful soldering requires that you follow some simple steps and get a lot of practice. It's important to remember that timing is critical when it comes to the art of soldering. As you read through the soldering procedure steps, pay close attention to words like "immediately" and "a few seconds" — and interpret them literally. Here are the steps for soldering a joint:

1. **Clean the metal surfaces to be soldered.**

   Wipe leads, wire ends, or etched circuit board surfaces (described later in this chapter) with isopropyl alcohol so that the solder adheres better. Let surfaces dry thoroughly before soldering: You don't want them to catch on fire!

2. **Secure the items being joined.**

   You can use a *third hand* clamp (as described in Chapter 9) or a vise and an alligator slip to hold a discrete component steady as you solder a wire to it, or use needlenose pliers to hold a component in place over a circuit board.

3. **Position the soldering iron.**

   Holding the iron like a pen, positioning the tip at a 30- to 45-degree angle to the work surface. (See Figure 11-4.)

4. **Apply the tip to the joint you're working on (as shown in Figure 11-4).**

   Do not apply heat directly to the solder. Allow the metal a few seconds to heat up.

5. **Feed the cold solder to the heated metal area.**

   The solder will melt and flow around the joint within a couple of seconds.

6. **Immediately remove the solder, and then remove the iron.**

   As you remove them, hold the component still until the solder cools and the joint solidifies.

7. **Place the soldering iron securely in its stand.**

   Never place a hot soldering iron on your work surface.

Be careful to use just the right amount of solder (which means applying solder for just the right amount of time): Use too little, and you form a weak connection; use too much, and the solder may form globs that can cause short circuits.

You can damage many electronic components if you expose them to prolonged or excessive heat, so take care to apply the soldering iron only long enough to heat a component lead for proper soldering — no more, no less.

To avoid damaging heat-sensitive components (such as transistors), attach an alligator clip to the lead between the intended solder joint and the body of the component. This way, any excess heat will be drawn away through the clip and will not damage the component. Be sure to let the clip cool before you handle it again.

Figure 11-4:
Holding the iron at an angle, apply heat to the metal parts you are soldering, and then feed the cold solder into the joint.

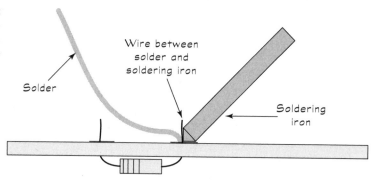

Inspecting the joint

After soldering, you should inspect the joint visually to make sure it's strong and conductive. The cooled solder joint should be shiny, not dull, and should be able to withstand a gentle tug from one side. If you soldered a lead to a circuit board, you should be able to see a *fillet* (volcano-shaped raised area of solder) at the joint. If you see dull solder or jagged peaks, you know you have a *cold solder joint.* Cold joints are physically weaker than properly made joints, and they don't conduct electricity as well.

Cold solder joints can form if you move the component while the solder is still cooling, if the joint is dirty or oily, or if you failed to heat the solder properly. Resoldering without first desoldering often produces cold solder joints because the original solder isn't heated enough.

If you have a cold solder joint, it's best to completely remove the existing solder (as described in the next section, "Desoldering when necessary"), clean the surfaces with isopropyl alcohol, and reapply fresh solder.

Desoldering when necessary

At some point in your work with electronics, you're bound to run into a cold solder joint, a backward-oriented component, or some other soldering mishap. To correct these mistakes, you have to remove the solder at the joint, and then apply new solder. You can use a desoldering pump (also known as a solder sucker and described shortly), a solder wick, or both to remove solder from the joint.

Use *solder wick* (also called *solder braid*), which is a flat braid of copper, to remove hard-to-reach solder. You place it over unwanted solder and apply heat. When the solder reaches its melting point, it adheres to the copper wire, which you then remove and dispose.

Exercise care when using solder wick. If you touch the hot braid, you can get a serious burn — copper conducts heat extremely well.

A *desoldering pump,* also known as a *solder sucker*), uses a vacuum to suck up excess solder that you melt with your soldering iron. There are two types:

- ✔ **A spring-loaded plunger-style pump:** To use a spring-loaded pump, you depress the plunger and position the nozzle over the joint that you want to remove. Then you carefully position the soldering iron tip into the joint to heat the solder, avoiding contact with the end of the pump. As the solder begins to flow, you release the plunger to suck up the solder. Finally, you expel the solder from the pump (into a waste receptacle) by depressing the plunger one more time. Repeat these steps as needed to remove as much of the old solder as possible.

  Don't store a desoldering pump with a cocked plunger. The rubber seal can become deformed, diminishing the vacuum to the extent that the pump may be unable to suck up any solder.

- ✔ **A bulb-style pump:** A bulb desoldering pump works a lot like the spring-loaded variety, except you squeeze the bulb to create the vacuum and release it to suck up the solder. You may find it difficult to use this pump unless you mount the bulb on the soldering iron. In fact, a device called a *desoldering iron* consists of a soldering iron with a vacuum bulb piggy-backed alongside it.

Cooling down after soldering

Make it a habit to unplug — not just shut off — your soldering iron when you finish your soldering work. Brush the tip of the still-warm iron against a damp sponge to clean off excess solder. After the iron has cooled down, you can use tip-cleaner paste to remove stubborn dirt. Then you can finish up with three good practices:

✔ Make sure the iron is completely cool before storing it.

✔ Place your solder reel in a plastic bag to keep it from getting dirty.

✔ Always wash your hands when you finish soldering, because most solder contains lead, which is poisonous.

Practicing safe soldering

Even if you plan to solder just one connection, you should take the appropriate precautions to protect yourself — and those around you. Remember, the iron will reach temperatures exceeding 700°F, and most solder contains poisonous lead. You (or a nearby friend or pet) may unwittingly be hit by popping, sputtering solder if you run across the occasional air pocket or other impurity in your reel. Just one small drop of solder hitting you in the eye, or a tumbling soldering iron that lands on your bare foot, can ruin a day, a body part — and a friendship.

Set up your work area — and yourself — with soldering safety in mind. Make sure the room is well ventilated, that you have the iron placed snugly in its stand, and that you position the electrical cord to avoid snags. Wear shoes (no flip-flops!), safety glasses, and an anti-static wristband when soldering. Avoid bringing your face too close to hot solder, which can irritate your respiratory system and possibly sputter. Keep your face to one side, and use a magnifying glass, if needed, to see tiny components that you are soldering.

Never solder a circuit with power flowing through it! Make sure the battery or other power supply is disconnected before you apply your soldering iron to components. If your iron has an adjustable temperature control, dial up the recommended setting for the solder you're using. And if your soldering iron accidentally goes belly-up, *stand back* and let it fall. If you try to grab it, Murphy's Law says you'll grab the hot end.

Finally, always unplug your soldering iron when you're done and promptly wash your hands.

Making a Commitment: Creating a Permanent Circuit

So you've perfected the world's greatest circuit, and you want to make it permanent. You have a few possible ways to transfer your circuit and make connections that last. Solder breadboards and perf boards are two popular foundations for permanent circuits. You can solder your connections onto either surface, but if you want additional flexibility, you can use a technique called wire wrapping to make connections on a perf board. This section covers the various methods of making a commitment to your circuit.

Making circuit boards with plug-and-play ICs

When you build circuit boards that include integrated circuits, instead of soldering the IC directly onto the board, use an IC socket. You solder the socket onto the board, and then after you're done soldering, you plug in the IC and hit the switch.

IC sockets come in different shapes and sizes, to match the integrated circuits they're meant to work with. For example, if you have a 16-pin integrated circuit, choose a 16-pin socket.

Here are some good reasons for using sockets:

✔ **Soldering a circuit board can generate static.** By soldering to the socket rather than to the actual IC, you can avoid ruining CMOS or other static-sensitive ICs.

✔ **ICs are often among the first things to go bad when you're experimenting with electronics.** Having the option to pull out a chip that you suspect is bad so you can replace it with a working one makes troubleshooting a whole lot easier.

✔ **You can share an expensive IC, such as a microcontroller, among several circuits.** Just pull the part out of one socket and plug it into another.

Sockets are available in all sizes to match the different pin arrangements of integrated circuits. They don't cost much — just a couple of pennies for each socket.

Moving your circuit to a solder breadboard

The *solder breadboard,* also called a *solder board,* an *experimenter's PC board,* or a *universal solder board,* allows you to take any design that you create on a solderless breadboard and make it permanent. You can transfer your design to a solder board easily because the solder breadboard has the exact same layout as the solderless breadboard.

To transfer your design, you simply pick the parts off your solderless breadboard, insert them in the solder breadboard, and solder them into place in the corresponding spots. Use wires in the same way you did in the original solderless breadboard: to connect components that aren't already electrically connected by the metal strips of the circuit board.

 If you design a really small circuit, you can use just half of a solder breadboard. Before transferring the components, cut the solder breadboard with a hack saw. Wear a protective mask to avoid breathing in the dust produced by the saw. Clean the portion of the board that you want to use and solder away.

Leave space at the corners of the board so you can drill mounting holes. You use these holes to secure the board inside whatever enclosure your project provides (such as the chassis of a robot). Alternatively, you can secure the board to a frame or within an enclosure by using double-sided foam tape. The tape cushions the board and prevents breakage, and the thickness of the foam prevents the underside of the board from actually touching the chassis.

Solder breadboards have one main disadvantage: They don't use space very efficiently. Unless you cram the components onto the board, the breadboard limits you to building circuits with only two or four integrated circuits and a handful of discrete components. In time, after some trial and error, you can figure out how to conserve space on a solder board.

Prototyping with pre-drilled perf boards

Another kind of general-purpose circuit board that you can use for your projects is a pre-drilled perf board. These boards go by many names, such as a grid board or a universal, general-purpose, or prototyping PC board. Perf boards come in a variety of sizes and styles, such as those shown in Figure 11-5. Some are bare, but most contain copper pads and traces for wiring. All styles are designed for you to use with ICs and other modern-day electronics components, which means the holes are spaced 0.1-inch apart.

You can choose the style of grid board according to the type of circuit you're building. Some grid layouts suit certain applications better than others. Personally, we prefer the plain universal PC board with interleaved buses, because we think they are easier to use. (You don't ride the bus on a circuit board, you solder things to it. A *bus* in the electronics world is just a common connection point.) You tie components together on the universal PC board, using three- (or more) point contacts.

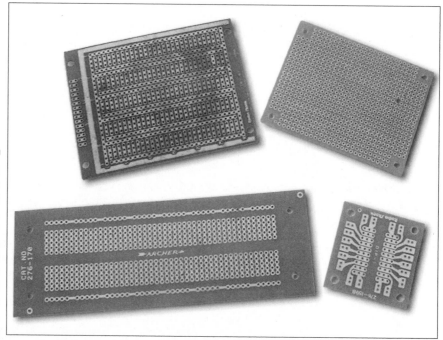

Figure 11-5:
A few sample perf boards, ready for you to clean them if necessary and add electronics components.

A bus runs throughout the circuit board so that you can easily attach components to it. Many perf boards have at least two buses, one for power and one for ground. The buses run up and down the board, as you can see in Figure 11-6. This layout works ideally for circuits that use many integrated circuits. Alternating the buses for the power supply and ground also helps to reduce undesirable inductive and capacitive effects.

To construct circuits using bare perf boards, which don't have coppers pads and traces for wire, you use the wire-wrapping method (discussed in the next section). Most perf boards contain pads and traces, so you can solder components directly onto them or use wire wrapping. You can use these perf board just as you use a solder breadboard. After cleaning the board so that the copper pads and traces are bright and shiny, plug the components into the board and solder them into place. Use insulated wire to connect components that aren't adjacent to one another.

Getting your wires wrapped

Wire wrapping is a point-to-point wiring system that uses a special tool and extra-fine 28- or 30-gauge wrapping wire. When you do it properly, wire-wrapped circuits are as sturdy as soldered circuits. And you have the added benefit of being able to make modifications and corrections without the hassle of desoldering and resoldering.

You have to limit wire wrapping to projects that use only low-voltage DC. It's not for anything that requires a lot of current, because the wire you use isn't large enough to carry much current.

To wire-wrap, you need:

- **Perf board:** You attach the components to this board. You can use a bare (no copper) board or one that has component pads for soldering. We personally prefer the padded board.
- **Wire-wrap sockets for ICs and other parts:** These sockets have extra-long metal posts. You wrap the wire around these posts.
- **Tie posts:** These posts serve as common connection points for attaching components together.
- **Wire-wrap wire:** The wire comes pre-cut or in spools. We prefer pre-cut wire, but try both before you form an opinion.
- **Wire-wrapping tool:** You can use this specific tool to wrap wire around a post and remove it. The tool also includes an insulation stripper; use this tool, not a regular wire stripper, to remove the insulation from wire-wrap wire.

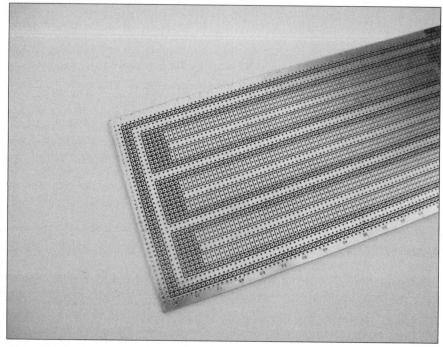

Figure 11-6:
Several
buses run
up and
down this
perf board.

Though you can wire-wrap directly to the leads attached to resistors, capacitors, diodes, and other components, most people prefer using wire-wrap sockets. The reason? Most components have round leads. A wire-wrap socket has square posts. The square shape helps to bite into the wire, keeping things in place. If you wrap directly to component leads, you may want to tack on a little bit of solder to keep the wire in place.

Wire wrapping is a straightforward process. You just insert all sockets into the perf board, use the wire-wrap tool to connect the sockets together, and then plug the ICs and other components into their sockets. If your perf board has solder pads, then you would be wise to touch a little solder between one of the pads and the post sticking through it. This dab of solder keeps the sockets from coming out.

A big advantage of wire-wrapping over soldering is that you can make changes relatively easily. Simply unwrap the wire and reroute it to another post. If the wire gets cruddy, just replace it with a new one.

There's more to wire-wrapping than this book can cover here. If it sounds like a method you think would be useful to you, do an Internet search using the following keywords: *wire wrapping technique electronics.* You can find numerous Web sites that help you become an expert wire-wrapper-upper, such as http://www.ee.ryerson.ca:8080/~jkoch/prototype/Proto.htm.

Making a custom circuit board

After you become experienced in the ways of designing and building electronics projects, you may want to graduate to the big time and create your own custom circuit board, geared for a particular circuit design. You can make (yes, make) your very own *printed circuit board (PCB)* — a surface that supports your circuit and includes interconnections along the surface of the board — just like the electronics manufacturers do. PCBs are reliable, rugged, allow for higher-density circuits, and enable you to include non-standard size components that may not fit in other types of circuit boards.

Making a printed circuit board is a fairly involved process — and beyond the scope of this book — but here's the lowdown on some of the steps involved:

1. You make a blank PCB by gluing or laminating a thin sheet of copper, known as *cladding,* onto the surface of a plastic, epoxy, or phenolic base. This forms a sort of "blank canvas" for the creation of a circuit.

2. You prepare a mask of your circuit layout, transfer it onto clear transparent film, and use the mask to expose a sheet of sensitized copper to strong ultraviolet light.

3. You dip the sensitized, exposed sheet into a developer chemical, producing a pattern (called a *resist pattern*) of the circuit board layout.

4. You form the circuit layout by etching away the portions of the copper not protected by the resistor — leaving behind just the printed circuit design, which consists of *pads* (contact points for components) and *traces* (interconnects).

5. You drill holes into the center of each pad so you can mount components on the top of the board with leads poking through the holes.

6. Finally, you solder each component lead to the board's pads.

To find out exactly how to make your own PCB, you can search the Internet with the following keywords: *make printed circuit board* — and you'll find tutorials, illustrations, and even videos that explain the process in great detail.

Chapter 12

Measuring and Analyzing Circuits

. . . .

In This Chapter

▶ Introducing your new best friend: your multimeter

▶ Using a multimeter to measure all kinds of things

▶ Setting up and calibrating your multimeter

▶ Making sure electronic components are working properly

▶ Probing around your circuits

▶ Identifying the cause of circuit problems

. . . .

*Y*our excitement builds as you put the finishing touches on your circuit. With close friends standing beside you, eager to witness the first of your ingenious electronics exploits, you hold your breath as you flip the power switch, and . . .

Nothing. At least, nothing at first. Then, disappointment, disillusionment, and disbelief as your friends — and your confidence — slowly retreat from the scene.

What could possibly be wrong? you ask yourself. Then you notice it: There's smoke emanating from what used to be a resistor. And then you realize you used a 10 Ω resistor instead of a 10kΩ resistor, trusting your old eyes and your worn-out mind to read and interpret resistor stripes properly. Oops!

In this chapter, you find out how to use an incredibly versatile tool — the multimeter — to perform important face-saving checks on electronic circuits and components. These tests help you determine whether everything is A-OK before you go showing off your circuitry to friends and family. When you're done reading this chapter, you will realize that your multimeter is as important to you as an oxygen tank is to a scuba diver: You can both get along okay on your own for a while, but sooner or later, you're bound to suffer unless you get some help.

Multitasking with a Multimeter

Because you can't follow the electrons in your circuits with your bare eyes — and you *don't* want to follow them with your bare hands — you need help from a versatile electronics test instrument known as a multimeter.

A *multimeter* is an inexpensive handheld testing device that can measure voltage, current, and resistance. Some can also test diodes, capacitors, and transistors. With this one handy tool, you can verify proper voltages, test whether you have a short circuit, determine whether there's a break in a wire or connection, and much more. Make friends with your multimeter, because it can help you make sure your circuits work properly and is an invaluable tool for scouting out circuit problems.

Figure 12-1 shows a typical multimeter. You turn a dial to select (a) the type of measurement you want to make and (b) a range of values for that measurement. You then apply the metal tips of the two test leads (one red, one black) to a component or some part of your circuit, and the multimeter displays the resulting measurement.

Multimeter test leads have conical tips that you hold in contact with the component you're testing. You can purchase special spring-loaded test clips that slip over the tips, making it easier for you to attach the test leads to the component leads or other wires. These insulated test clips ensure a good connection between the test leads and whatever it is you're testing, while preventing accidental contact with another part of the circuit.

It's a voltmeter!

Multimeters can measure both DC and AC voltages. They provide a variety of voltage measurement ranges, from 0 volts to a maximum voltage. A typical set of DC voltage ranges is 0–0.25 V, 0–2.5 V, 0–10 V, 0–50 V, and 0–250 V. You can use the multimeter as a *voltmeter* to measure the voltage of a battery outside of a circuit or *under load* (meaning when it is providing power within a circuit). You can also use it with your circuit powered up if you want to test voltages dropped across circuit elements and (for that matter) voltages at various points in your circuit with respect to ground.

Voltmeters are so important in electronics, they even have their very own circuit symbol, shown here. You may see this symbol with leads touching points in a circuit you read about on a Web site or in an electronics book. It tells you to take a voltage measurement across the two indicated points.

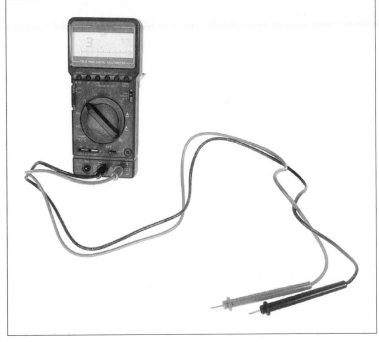

Figure 12-1:
Multimeters
measure
voltage,
resistance,
current, and
continuity.

Your multimeter can help you pinpoint the location of a problem in your circuit. It can verify whether the proper voltage reaches a component, such as a light-emitting diode (LED) or a switch. You use multimeter tests to narrow down the field of suspects until you find the culprit causing all your headaches.

It's an ammeter!

Your multimeter also functions as an *ammeter,* a device that measures the current going through a circuit. You use this function of the multimeter to determine whether a circuit or component is drawing too much current. If your circuit has more current going through it than it's designed to handle, the components may get overheated and damage your circuit permanently.

 The circuit symbol for an ammeter is shown here.

Ohm my! It's an ohmmeter, too!

You can measure the resistance of an individual component or a entire circuit (measured in ohms, as detailed in Chapter 3) with your multimeter functioning as an *ohmmeter*. You use this function to check up on wires, resistors, motors, and many other components. You always test resistance with the circuit *unpowered*. If the circuit is powered, current flowing through it can invalidate the resistance readings — or damage the meter.

 The circuit symbol for an ohmmeter is shown here.

 If you're measuring the resistance of an individual component, take it out of the circuit before you test it. If you test a resistor when it's wired in to a circuit, you'll get the equivalent resistance between two points, which is not necessarily the resistance of just your resistor. (See Chapter 3 for more on equivalent resistance.)

Because resistance, or (for that matter) lack of resistance, can reveal short circuits and open circuits, you can use your ohmmeter to sniff out problems such as breaks in wires and hidden shorts between components. A short circuit generates an ohmmeter reading of zero (or virtually zero) resistance; an open circuit generates a readout of infinite resistance. If you test the resistance from one end of a wire to the other and you get an infinite readout, then you know there must be a break somewhere along the length of the wire. Such tests are known as *continuity tests*.

By measuring resistance, you can tell whether the following circuit elements and connections are working properly:

- **Fuses:** A blown fuse generates an infinite resistance reading, indicating an open circuit.

- **Switches:** An "on" switch should generate a zero (or low) resistance reading; an "off" switch should generate an infinite reading.

- **Circuit board traces:** A bad copper trace (line) on a printed circuit board acts like a broken wire and generates an infinite resistance reading.

- **Solder joints:** A bad joint may generate an infinite resistance reading.

 Many multimeters include an audible continuity-testing feature. By turning the meter's selector to Continuity or Tone, you can hear a beep whenever the meter detects continuity in a wire or connection. If the wire or connection doesn't have continuity, the meter stays silent. The audible tone gives you a convenient way to check an entire circuit without having to keep your eye on the meter.

Exploring Multimeters

Multimeters range from bare-bones handheld models that cost about $10 to feature-rich hobbyist models that cost anywhere from $50 to over $100, to sophisticated industrial bench-top models that cost thousands of dollars. Even a low-end multimeter can really help you understand what's going on in low-voltage circuits — but unless you're really cash-strapped, it's a good idea to spend a little more on a multimeter to get more features; you're sure to find them useful as you expand your electronics horizons.

Choosing a style: analog or digital

Most multimeters today, including the one shown in Figure 12-1, are *digital multimeters,* provide readouts on a digital (numeric) display. You may also find some older-style *analog multimeters* that use a needle to point to a set of graduated scales. You can see an analog multimeter in Figure 12-2.

An analog multimeter can be a bit challenging to use. After selecting the type of testing (voltage, current, or resistance) and the range, you must correlate the results by using the appropriate scale on the meter face, and estimate the reading as the needle swings into action. It's easy to get an erroneous reading — whether due to misinterpreted scale divisions, mental arithmetic errors, or a compromised view of where the needle is pointing. In addition, resistance measurements are rather imprecise because the measurement scale is compressed at high resistance values.

Digital multimeters display each measurement result as a precise number, taking the guesswork out of the reading process. Most handheld digital multimeters are accurate to within 0.8% for DC voltages; the pricey bench-top varieties are over 50 times more accurate. Many digital multimeters also include an *auto-ranging* feature, which means that the meter automatically adjusts itself to display the most accurate result possible; some have special testing features for checking diodes, capacitors, and transistors.

Analog multimeters actually outperform digital multimeters when it comes to detecting *changing* readings, but if you don't have much of a need for that, your best bet is to get a digital multimeter because of its ease of use and more accurate readings.

Figure 12-2:
An analog mul-timeter uses a needle to indicate voltage, current, and other values.

Taking a closer look at a digital multimeter

All digital multimeters perform the basic voltage, resistance, and current measurements. Where they differ is in the range of values they can measure, the additional measurements they can perform, the resolution and sensitivity of their measurements, and the extra bells and whistles they come with.

Be sure to at least browse through the manual for the multimeter you purchase. It contains a description of the features and specifications for your meter, as well as important safety precautions.

Here's what you'll find when you explore a digital multimeter:

✔ **Power switch/battery/fuse:** The on/off switch connects and disconnects the battery that powers the multimeter. Many multimeters use standard-size batteries, such as a 9-volt or AA cell, but pocket-size meters use a coin-type battery. (Avoid using rechargeable batteries in a multimeter; they may produce erroneous results for some models.) Most multimeters use an internal fuse to protect themselves against excessive current or voltage; some come with a spare fuse (if yours doesn't, buy one).

- **Function selector:** Dial this knob to choose a test to perform (voltage, current, resistance, or something else) and, on some models, the range setting you want to use. Some multimeters are more "multi" than others, and include one or more of the following categories: AC amperes, capacitance, transistor gain (h_{FE}), and diode test. Many multimeters further divide some measurement categories into 3–6 different ranges; the smaller the range, the greater the sensitivity of the reading. Figure 12-3 shows a close-up of a function selector dial.

- **Test leads and receptacles:** Inexpensive multimeters come with basic test leads, but you can purchase higher-quality coiled leads that stretch out to several feet and recoil to a manageable length when not in use. You may also want to purchase spring-loaded clip leads that stay in place and are insulated to prevent the metal tip from coming into contact with other parts of your circuit. Some multimeters with removable test leads provide more than two receptacles for the leads. You insert the black test lead into the receptacle labeled GROUND or COM, but the red lead may be inserted into a different receptacle depending on what function and range you've dialed up. Most meters provide additional input sockets for testing capacitors and transistors, as shown in the top-right corner of Figure 12-3. Refer to your multimeter manual for details.

Figure 12-3:
Digital
multimeters
provide a
wide variety
of measure-
ment options.

✔ **Digital display:** The readout is given in units specified by the range you've dialed up. For instance, a reading of 15.2 means 15.2 V if you've dialed a 20 V range, or 15.2 millivolts (mV) if you've dialed a 200 mV range. Most digital multimeters designed for hobbyists have what's called a *3¹/₂ digit display:* Its readout contains three or four digits, where each of the three right-most digits can be any digit from 0 to 9, but the optional fourth digit (that is, the left-most — most significant — digit) is limited to 0 or 1. For instance, if set to a 200 V range, such a multimeter can give readouts ranging from 00.0 V to 199.9 V.

Homing in on the range

Many digital multimeters (and most analog multimeters) require that you select the range before the meter can make an accurate measurement. For example, if you're measuring the voltage of a 9 V battery, you set the range to the setting closest to (but still above) 9 volts. For most meters, this means you select the 20 V or 50 V range.

If you select too big a range, the reading you get won't be as accurate. (For instance, on a 20 V range setting, your 9 V battery may produce a reading of 8.27 V, but on a 200 V range setting, the same battery produces a reading of 8.3 V. You often need as much precision in your readings as possible.)

If you select too small a range, a digital multimeter typically displays a flashing 1 (or OL), whereas the needle on an analog meter shoots off the scale, possibly damaging the precision needle movement (so make sure you start with a large range and dial it down, if necessary). If you see an *over range* indicator when testing continuity, it means the resistance is so high that the meter can't register it; it's fairly safe to assume that's an open circuit.

The auto-ranging feature found on many digital multimeters makes it even easier to get a precise reading. For instance, when you want to measure voltage, you set the meter function to Volts (either DC or AC) and take the measurement. The meter automatically selects the range that produces the most precise reading. If you see an over range indicator (a flashing 1 or OL), that's telling you the value is too high to be measured by the meter. Auto-ranging meters don't require range settings, so their dials are a lot simpler.

There's a limit to what a multimeter can test. You call that limit its *maximum range.* Most consumer multimeters have roughly the same maximum range for voltage, current, and resistance. For your hobby electronics, any meter with the following maximum ranges (or better) should work just fine:

DC volts: 1,000 V

AC volts: 500 V

DC current: 200 mA (milliamperes)

Resistance: 2MΩ (two megohms, or 2 million ohms)

What if you need to test higher currents?

Most digital multimeters limit current measurements to less than one amp. The typical digital multimeter has a maximum range of 200 milliamperes. Attempting to measure substantially higher currents may cause the fuse in the meter to blow. Many analog meters, especially older ones, support current readings of 5 or 10 amps, maximum.

You may find analog meters that can tolerate a high-ampere input handy if you're testing motors and circuits that draw a lot of current. If you have only a digital meter with a limited-milliampere input, you can still measure higher currents indirectly by using a low-resistance, high-wattage resistor. To do this, you place a 1 Ω, 10-watt resistor in series with your circuit so that the current you want to measure passes through this test resistor. Then you use your multimeter as a voltmeter, measuring the voltage dropped across the 1 Ω resistor. Finally, you apply Ohm's Law to calculate the current flowing through the test resistor as follows:

$$current = {}^V\!/_R = {}^V\!/_1\,\Omega$$

Because the nominal value of the resistor is 1 Ω, the current (in amps) through the resistor has roughly the same value as the voltage (in volts) you measure across the resistor. Note that the resistor value will not be exactly 1 Ω in practice, so your reading may be off as much as 5%–10%, depending on the tolerance of your resistor and the accuracy of your meter. (You can get a refresher on Ohm's Law in Chapter 3.)

Setting Up Your Multimeter

Before testing your circuits, you must make sure that your meter is working properly. Any malfunction gives you incorrect testing results — and you may not even realize it. To test your multimeter, follow these steps:

1. **Make sure that the test probes at the end of the test leads are clean.**

   Dirty or corroded test probes can cause inaccurate results. Use electronic contact cleaner to clean both ends of the test probes and, if necessary, the connectors on the meter.

2. **Turn on the meter and dial it to the Ohms (Ω) setting.**

   If the meter isn't auto-ranging, set it to low ohms.

3. **Plug both test probes into the proper connectors of the meter and then touch the ends of the two probes together (see Figure 12-4).**

   Avoid touching the ends of the metal test probes with your fingers while you're performing the meter test. The natural resistance of your body can throw off the accuracy of the meter.

4. **The meter should read 0 (zero) ohms or very close to it.**

If your meter doesn't have an auto-zero feature, press the Adjust (or Zero Adjust) button. On analog meters, rotate the Zero Adjust knob until the needle reads 0 (zero). Keep the test probes in contact and wait a second or two for the meter to set itself to zero.

5. **If you don't get any response at all from the meter when you touch the test probes together, recheck the dial setting of the meter.**

   Nothing happens if you have the meter set to register voltage or current. If you make sure that the meter has the right settings and it still doesn't respond, you may have faulty test leads. If necessary, repair or replace any bad test leads with a new set.

Figure 12-4:
Touch the test probes of the meter together and verify a zero ohms reading to make sure the meter is working properly.

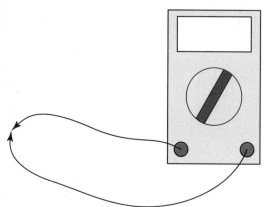

You can consider the meter *calibrated* when it reads zero ohms with the test probes shorted together (held together so they're touching each other). Do this test each time you use your meter, especially if you turn off the meter between tests.

If your meter has a Continuity setting, don't use it to zero-adjust (calibrate) the meter. The tone may sound when the meter reads a few ohms, so it doesn't give you the accuracy you need. Recalibrate the multimeter using the Ohms setting, and not the Continuity setting, to ensure proper operation.

Operating Your Multimeter

When you use your multimeter to test and analyze circuits, you must consider what settings to dial up, whether you're testing components individually or as part of a circuit, whether the circuit should be powered up or not, and where you place the test leads (in series or in parallel with whatever you're testing).

Think of your multimeter as an electronic component in your circuit (because in a way, it is). If you want to measure voltage, your meter must be placed *in parallel* with the section of the circuit you're measuring, because voltages across parallel branches of a circuit are the same. If you want to measure current, your meter must be placed *in series* with the section of the circuit you are measuring, because components in a series circuit carry the same current. (You can read about series and parallel connections in Chapter 2.)

The following sections explain exactly how to measure voltage, current, and resistance using a multimeter.

Measuring voltage

To examine voltage levels — that is, the voltage drop from a point in your circuit to ground — throughout your circuit using a multimeter, your circuit must be powered up. You can test the voltage at almost any point in a circuit, not just the battery connections. Here's how to test voltages:

1. **Set up your meter as described in the earlier section "Setting Up Your Multimeter."**

2. **Dial up the type of voltage (AC or DC).**

3. **Choose the range that gives you the most sensitivity.**

4. **Attach the black lead of the meter to the ground connection of the circuit.**

5. **Attach the red lead of the meter to the point in the circuit that you want to measure.**

   This places your multimeter in parallel with the voltage drop you want to measure.

Figure 12-5 shows an example of using a multimeter to measure the voltage at a couple of different points in a simple 555 astable circuit (described in Chapter 7), which produces a series of voltage pulses that automatically alternates between low (0 V) and high (the positive supply voltage). The multimeter is set to DC volts with a range of 0–20 V. In the top image, the meter is measuring the voltage that powers the entire circuit, so the multimeter reads 9 V. In the bottom image, the meter is measuring the voltage at the output of the 555 timer IC. Because the output of the 555 timer IC alternates between low and high, the meter reading alternates between 0 V and 9 V (the positive supply voltage).

Depending on the resistor and capacitor values in the 555 timer circuits shown in Figure 12-5, the output may change so rapidly that your multimeter may not be able to keep up with the voltage swings. (Chapter 7 explains how this works.) To test fast-changing signals, you need a logic probe (for digital signals only) or an oscilloscope. (More about both of these test instruments in Chapter 13.)

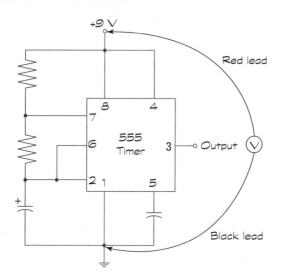

Testing Supply Voltage

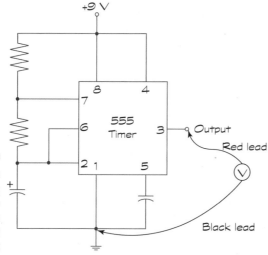

Figure 12-5:
Measuring
two differ-
ent voltages
on a 555
timer IC
circuit.

Testing Output Voltage

Measuring current

When you use the multimeter as an ammeter, you connect the meter *in series with* the component you want to measure current through, so that the exact same current you want to measure passes through the meter. This setup, as shown in Figure 12-6, is very different from the voltmeter configuration:

1. **Set up your meter as described in the earlier section "Setting Up Your Multimeter."**

2. **Dial up the type of current (AC or DC).**

3. **Choose the range that gives you the most sensitivity.**

4. **Interrupt the circuit at the point where you want to measure current: Attach the black lead of the meter to the more negative side of the circuit and the red lead to the more positive side.**

   This places your multimeter in series with the component you want to measure current through.

   Make sure the circuit is powered up. If you don't get a reading, try reversing the connections of the leads to the multimeter.

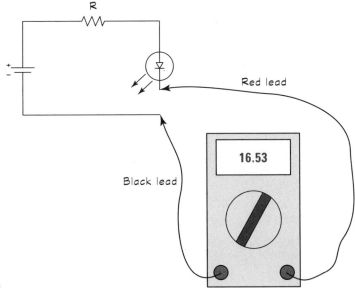

Figure 12-6:
Measuring
current
involves
connecting
the meter
in series
with the
circuit or
component.

To measure how much overall current an entire circuit draws, you insert your meter's leads in series with the positive power supply. But remember that many digital meters are limited to testing currents of 200 mA or less. Be careful: Don't test higher current if your meter isn't equipped to do so.

Never leave your multimeter in an ammeter position after measuring current. You can damage the meter. Get into the habit of turning the meter off immediately after running a current test.

Don't blow your fuse!

Many analog and digital meters provide a separate input (test lead receptacle) for testing current, usually marked as A (for amps) or mA (for milliamps). Some multimeters provide an additional input for testing higher currents, typically up to 10 amps. The multimeter shown in Figure 12-3 has two inputs for testing current, labeled mA and 10A.

Be sure to select the appropriate input *before* making any current measurement. Forgetting to do this step may either blow a fuse (if you're lucky) or damage your meter (if you're unlucky).

Measuring resistance

You can run lots of different tests using your multimeter as an ohmmeter that measures resistance. Obviously, you can test resistors to check their values or see whether they've been damaged, but you can also examine capacitors, transistors, diodes, switches, wires, and other components using your ohmmeter. Before you measure resistance, however, be sure to calibrate your ohmmeter (as described in the earlier section "Setting Up Your Multimeter").

If your multimeter has specific features for testing capacitors, diodes, or transistors, we recommend that you use those features rather than the methods that we give you in the following sections. But if you have a bare-bones multimeter without those features, these methods can really help you.

Testing resistors

Resistors are components that limit current through a circuit. (See Chapter 3 for details.) Sometimes you need to verify that the nominal resistance value marked on the body of a resistor is accurate, or you may want to investigate whether a suspicious-looking resistor with a bulging center and third-degree burns has gone bad.

To test a resistor with a multimeter, follow these steps:

1. **Turn the power off before you touch your circuit, and then disconnect the resistor you want to test.**

2. **Set your multimeter to read ohms.**

   If you don't have an auto-ranging meter, start at a high range and dial down the range as needed.

3. **Position the test leads on either side of the resistor.**

Take care not to let your fingers touch the metal tips of the test leads or the leads of the resistor. If you touch them, you add the resistance of your body into the reading, producing an inaccurate result.

The resistance reading should fall within the tolerance range of the nominal value marked on the resistor. For instance, if you test a resistor with a nominal value of 1kΩ and a tolerance of 10 percent, your test reading should fall in the range of 900 to 1,100 Ω. A bad resistor can be either completely open inside (in which case you may get a reading of infinite Ω), or it can be shorted out (in which case you get a reading of zero Ω).

Testing potentiometers

As with a resistor, you can test a *potentiometer* (also called a *pot*) — which is a variable resistor — using the Ohms setting on your multimeter. (For more about pots, see Chapter 3.) Here's how to do the test:

1. **Turn the power off before you touch your circuit, and then remove the potentiometer.**

2. **Position the test leads on two of the potentiometer's leads. Depending on where you put the leads, you can expect one of these results:**

   • With the meter leads applied to one fixed end (point 1) and the *wiper,* or variable lead (point 2) (as shown in Figure 12-7), turning the dial shaft in one direction increases the resistance, whereas turning the dial shaft in the other direction decreases the resistance.

   • With the meter leads applied to the wiper (point 2) and the other fixed end (point 3), the opposite resistance variation happens.

   • If you connect the meter leads to both fixed ends (points 1 and 3), the reading that you get should be the maximum resistance of the pot, no matter how you turn the dial shaft.

Figure 12-7: Connect the test leads to the first and center, center and third, and first and third terminals of the pot.

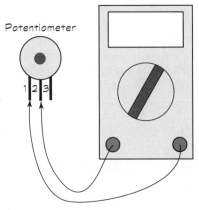

As you turn the shaft of the potentiometer, take note of any sudden changes in resistance, which may indicate a fault inside the pot. Should you find such a fault, replace the pot with a new one.

Testing capacitors

You use a *capacitor* to store electrical energy for a short period of time. (For coverage of capacitors, see Chapter 4.) If your multimeter doesn't have a capacitor-testing feature, you can still use it in the ohmmeter setting to help you decide whether to replace a capacitor. Here's how to test a capacitor:

1. **Before you test a capacitor, be sure you discharge it to clear all the electrical energy out of it.**

   Large capacitors can retain a charge for long periods of time — even after you remove power.

   To discharge a capacitor, you short out its terminals through an insulated *bleeder jumper* (as shown in Figure 12-8), which is simply a wire with a large (1MΩ or 2MΩ) resistor attached. The resistor prevents the capacitor from shorting out, which would make it unusable.

2. **After discharging the capacitor and removing the bleeder jumper, dial your multimeter to Ohms and touch the test leads to the capacitor leads.**

   Unpolarized capacitors don't care which way you connect the leads, but if you're testing a polarized capacitor, connect the black lead to the negative terminal of the capacitor, and the red lead to the positive terminal. (Chapter 4 explains how to determine capacitor polarity.)

3. **Wait a second or two, and then note the reading.**

   You'll get one of these results:

   - A good capacitor shows a reading of infinity when you perform this step.

   - A zero reading may mean that the capacitor is shorted out.

   - A reading of between zero and infinity could be indicative of a leaky capacitor, one that is losing its ability to hold a charge.

This test doesn't tell you whether the capacitor is open, which can happen if the component becomes structurally damaged inside or if its *dielectric* (insulating material) dries out or leaks. An open capacitor reads infinite ohms, just like a good capacitor. For a conclusive test, use a multimeter with a capacitor-testing function.

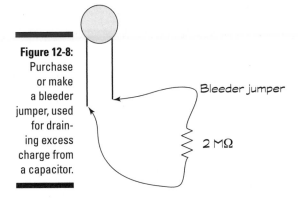

Figure 12-8:
Purchase
or make
a bleeder
jumper, used
for drain-
ing excess
charge from
a capacitor.

Bleeder jumper

2 MΩ

Testing diodes

A *diode* is a semiconductor component that acts like a one-way value for current. (To get the details on diodes, see Chapter 6.) If your multimeter doesn't
have a diode-check setting, you can use the Ohms setting to test most types
of diodes. To test a diode, follow these steps:

1. **Set your meter to a low-value resistance range.**

2. **Connect the black lead to the cathode (negative side, with a stripe)
   and the red lead to the anode (positive side).**

   The multimeter should display a low resistance.

3. **Then reverse the leads, and you should get an infinite resistance
   reading.**

If you're not sure which end is up in a diode you've got on hand, you can use
your multimeter to identify the anode and the cathode. Run resistance tests
with leads connected one way, and then with the leads connected the other
way. For the lower of the two resistance readings, the red lead is connected to
the anode and the black lead is connected to the cathode.

Testing transistors

A *bipolar transistor* is essentially two diodes in one package, as illustrated in
Figure 12-9. (For a PNP transistor, both diodes are reversed.) If your multimeter has neither a transistor-checking feature nor a diode-checking feature,
you can use the Ohms setting to test most bipolar transistors, in much the
same way you test diodes: You set your meter to a low-value resistance
range, and test each diode within your transistor in turn.

If you're testing an NPN transistor (such as the one shown in Figure 12-9), follow these steps:

1. **Set your meter to a low-value resistance range.**

2. **Connect the black lead to the collector of the transistor and the red lead to the base.**

   The multimeter should display a low resistance.

3. **Reverse the leads.**

   You should get an infinite resistance reading.

4. **Connect the black lead to the emitter and the red lead to the base.**

   The meter should display a low resistance.

5. **Reverse the leads.**

   The meter should display infinite resistance.

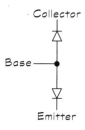

Figure 12-9:
A bipolar transistor is like two diodes in one package.

NPN transistor

Equivalent diode circuit

For a PNP transistor, the readings should be the opposite of what they are for an NPN transistor.

Use this test *only* with bipolar transistors. Testing with a multimeter can permanently damage some types of transistors, especially field-effect transistors (FETs). If you're not sure what type of transistor you have, look it up in a datasheet before you test it. You can often find the datasheet by searching the Internet for the component-identification number (for instance, search on "2n3906 datasheet").

Testing wires and cables

You can use your multimeter as an ohmmeter to run continuity tests on wires and cables. You may want to do this so you can sniff out breaks inside wires and *short circuits,* or unintended continuity, between two wires in a cable.

Even wire resists the flow of electrons

Why don't you *always* get 0 Ω when you test a wire, especially a long wire? All electrical circuits exhibit some resistance to the flow of current; the ohms measurement tests this resistance. Even short lengths of wire have resistance, but it's usually well below 1 Ω, so it's not an important test subject for continuity or shorts.

However, the longer the wire, the greater the resistance — especially if the wire has a small diameter. Usually, the larger the wire, the lower its resistance per foot. Even though the meter doesn't read exactly 0 Ω, you can assume proper continuity in this instance if you get a low ohms reading.

To test for continuity in a single wire, you connect the multimeter test leads to either end of the wire and dial up a low-range Ohms setting. You should get a reading of 0 Ω, or a very low number of ohms. A reading of more than just a few ohms indicates a possible break in along the wire, causing an *open circuit*.

To test for a short between different wires that shouldn't be electrically connected, you set the meter to measure ohms, and then connect one of the test leads to an exposed end of one wire and the other test lead to an exposed end of the other wire. If you get a reading of 0 Ω or a low number of ohms, you may have a short circuit between the wires. A higher reading usually means your wires are not shorted together. (Note that you may get a reading other than infinite ohms if the wires are still connected to your circuit when you make the measurement. Rest assured that your wires are not shorted unless your reading is very low or zero.)

Testing switches

Mechanical switches can get dirty and worn, or sometimes even break, making them unreliable or completely unable to pass electrical current. Chapter 8 describes four common types of switches: single-pole single-throw (SPST), single-pole double-throw (SPDT), double-pole single-throw (DPST), and double-pole double-throw (DPST). Depending on the switch, there may be zero, one, or two "off" positions, and there may be one or two "on" positions.

You can use your multimeter set to Ohms to test any of these switches. Be sure to familiarize yourself with the on/off position(s) and the terminal connections of the particular switch you're testing — and run tests for each possibility. With your test leads connected across the terminals of any input/output combination placed in the "off" position, the meter should read infinite ohms; the "on" position should give you a reading of 0 Ω.

You can most easily test switches by taking them out of a circuit. If the switch is still wired into a circuit, the meter may not show infinite ohms when you place the switch in the "off" position. If, instead, you get a reading of some value other than 0 Ω, you can assume the switch is operating properly as an open circuit when it's in the "off" position.

Testing fuses

Fuses are designed to protect electronic circuitry from damage caused by excessive current flow and, more importantly, to prevent a fire if a circuit overheats. A blown fuse is an open circuit that's no longer providing protection, so it has to be replaced. To test a fuse, set your multimeter to the Ohms setting and touch one test lead to either end of the fuse. If the meter reads infinite ohms, it means you have a burned-out fuse.

Running other multimeter tests

Many digital multimeters include extra functions that test specific components, such as capacitors, diodes, and transistors. These tests provide more definitive results than the resistance measurements we discuss earlier in this section.

If your multimeter has a capacitor-testing feature, it will display the value of the capacitor. This can really come in handy, because not all capacitors follow the industry standard identification scheme. Refer to your multimeter manual for the exact procedure because the specifics vary from model to model. Be sure to observe the proper polarity when connecting the capacitor to the test points on the meter.

If your multimeter has a diode-check feature, you can test a diode by attaching the red test lead to the anode (positive terminal) of the diode, and the black lead to the cathode (negative terminal). You should get a fairly low, but not zero, reading (for instance, 0.5). Then reverse the leads and you should get an over range reading. If you get two zero readings or two over-range readings, chances are your diode is bad (that is, shorted out or open).

You can use the diode-check feature to test bipolar junction transistors, treating them as two individual diodes, as shown in Figure 12-9.

If your multimeter has a transistor-checking feature, follow the procedure outlined in the manual, which varies from one model to another.

Using a Multimeter to Check Your Circuits

One of the top benefits of a multimeter is that it can help you analyze the rights and wrongs of your circuits. By using the various test settings, you can verify the viability of individual components and confirm that voltages and currents are what they should be. Inevitably, sooner or later, you hook up a circuit that doesn't work right away — but your multimeter can help you sniff out the problem if you can't resolve the problem by physically checking all your connections.

To troubleshoot your circuit, you should first mark up your circuit diagram with component values, estimated voltage levels at various points in the circuit, and expected current levels in each branch of the circuit. (Often the process of marking up the diagram uncovers a math error or two.) Then use your multimeter to probe around.

Here's a quick list of items to check as you troubleshoot your circuit:

- ✔ Check power supply voltages.
- ✔ Check individual component functionality and actual values (out of the circuit).
- ✔ Check continuity of wiring.
- ✔ Check voltage levels at various points in the circuit.
- ✔ Check current levels through part of the circuit (without exceeding the current capabilities of your multimeter).

Using a step-by-step procedure, you can test various components and parts of your circuit and narrow down the list of suspects until you either uncover the cause of your circuit problem or admit you need professional help — from your friendly neighborhood electronics guru.

Chapter 13

Getting Down with Logic Probes and Oscilloscopes

Chapter 12 talks about how to use a multimeter to test for all sorts of glitches and gotchas in your electronic circuits. Your meter is the most important tool on your workbench, but don't think it's the only thing that you can use to test your electronics stuff. If you're really, really serious about electronics, you may want to get several other testing tools for your workbench.

This chapter introduces you to two additional tools — a logic probe and an oscilloscope — and to how they can make you a more effective electronics troubleshooter. Neither tool is a "must-have" when you're just starting out, but as you work your way into intermediate and advanced electronics, consider adding them to your electronics workbench.

Probing the Depths of Logic

A *logic probe* like the one shown in Figure 13-1 — a fairly inexpensive tool — is specifically designed to test digital circuits, which handle just two voltage levels:

▶ **Low** (zero V or thereabouts), indicating logical 0 (zero).

▶ **High** (12 V or less, but most commonly 5 V), indicating logical 1 (one).

(You can read all about digital logic in Chapter 7.) The logic probe simply checks for high or low signals and turns on a different LED indicator depending on the signal it detects. A third LED indicator glows when the logic probe detects *pulsing,* which is when a signal alternates between high and low very quickly. Most logic probes also include a tone feature, so you can hear highs and lows while keeping your eyes on your circuit.

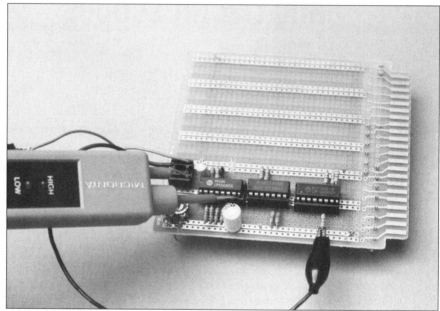

Figure 13-1:
The logic probe is useful for troubleshooting digital circuits.

Of course, you can always use your versatile multimeter to test digital circuits — you just have to translate voltage readings into logic states — but there is one major advantage of using a logic probe instead. If your circuit is bad and there is no signal at the point you're testing, your multimeter may give you a reading of zero volts, which you might interpret as a low signal (logical 0). If you use a logic probe on a bad circuit, and there is no signal, none of the probe lights glow and the probe doesn't make a sound. (This could, however, indicate that your probe is incorrectly connected.)

Logic probes receive power from the circuit under test. Most probes work with a minimum supply voltage of about 3 V and a maximum of no more than 15 V (sometimes more, sometimes less). Check the manual for the exact voltage range of your logic probe, and be sure to determine whether the supply voltage of your circuit falls within the acceptable range for the logic probe. Voltages exceeding the maximum can damage the logic probe.

Be extra careful about safety when using a logic probe. It's an active-circuit tester, designed to operate on circuits that are powered up. Keep your fingers away from the metal at the end of the probe and test leads. Take special care if you're testing a circuit (such as a dysfunctional DVD player) that runs off AC power and you need to expose the power-supply components to perform the test. Always consider that you may expose dangerously high voltages when you remove the cover from any AC-operated equipment; be sure to cover the equipment with insulating plastic to prevent accidental shock.

A logic probe has four connections, as you can see in Figure 13-2, three of which are leads with alligator clips and one of which is the probe itself. To use the probe, you need to make these four connections:

1. **Clip the black power lead to the circuit ground.**

2. **Clip the red power lead to the circuit voltage supply.**

   Be sure that this supply doesn't exceed the maximum voltage (usu-ally about 15 V) specified for your logic probe, or you may damage the probe.

3. **Clip the second black lead (ground) to the circuit ground.**

   This separate ground is important; if you fail to securely connect the probe to the circuit ground, the probe may yield erratic results — or not work at all.

4. **Place the tip of the probe against the part of the circuit that you're testing.**

To verify that you've connected the logic probe properly, you can touch the tip of the probe to the power supply of the circuit. The probe should indicate a high value. Then touch the tip of the probe to the circuit ground; the probe should indicate a low value. If either or both of these tests fail, examine the connection of the probe to the circuit and make corrections as necessary.

After you've made the connections, the probe's indicator lights and audible tones (on most models) help you determine the logic level at the test point:

- ✔ **A low indicator** (and low buzz tone) tells you that the test point has a logic low (at or about 0 V).

- ✔ **A high indicator** (and high buzz tone) tells you that the test point has a logic high (usually at or about 5 V).

- ✔ **Quickly toggling low and high indicators** mean that the logic signal is pulsing (changing quickly between low and high). Most probes have a separate indicator that tells you when a circuit is pulsing, and some also generate an intermittent buzz tone.

- ✔ **No indicator** tells you that the test point has no discernable high, low, or pulsing signal.

Why don't all circuits like logic probes?

Most test gear, including the multimeter and oscilloscope, draws very little current from the circuit that you're testing. Their makers design these testing tools this way so that the tools themselves don't influence the reading. Obviously, it does no good to test a circuit if the testing tool changes the behavior of that circuit. You can't get a reliable result.

Logic probes not only draw power from the circuit, but they also can load down the signal line that you're testing. The additional load of the logic probe may cause some of the weaker digital signals to drop in voltage to a point where you can't get an accurate reading.

Although this situation doesn't come up all that often, it's a good example of why you need to be somewhat familiar with the circuit that you're testing. Just know that poking the probe into unknown territory may yield unpredictable results.

Be sure to read the manual or instruction booklet that comes with your logic probe for additional pointers, cautions, caveats, warnings, and operating tips. Though many logic probes are similar in design, slight differences can influence the types of circuits that a particular probe works best with.

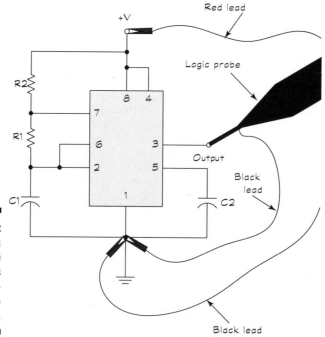

Figure 13-2:
The logic probe requires four connections to the circuit.

In addition to low and high output signals, some integrated logic circuits have a third state, called *Hi-Z,* or *high-impedance.* In simple terms, having this third state lets you connect a lot of outputs directly together, with only one being active (or enabled) at a time. The remaining outputs are set to their Hi-Z state, which makes them essentially invisible to the enabled output. The circuit only engages one output, either low or high, at any one time. The other outputs are put to sleep in the Hi-Z state and get activated in their own due time.

Scoping Out Signals with an Oscilloscope

For the average amateur electronics hobbyist working at home or at school, the oscilloscope is a nice tool to have around, but it's not absolutely necessary. An *oscilloscope* is a (rather expensive) piece of test equipment that displays how a voltage varies with time as a *trace* across a cathode-ray tube (CRT) or other display.

Serious electronics wizards may fork over a paycheck or two to purchase a bench oscilloscope like the one shown in Figure 13-3, but for about $200, you can purchase a battery-operated handheld scope with a liquid crystal display. You won't get the advanced features of the better bench scopes, but you'll still be able to view signal variations over time — something your multimeter simply cannot do. Another option is to spend several hundred dollars on a PC-based oscilloscope that uses your personal computer to store and display the electrical signals you measure. Most PC-based scopes are self-contained in a small, external module that connects to the desktop PC or laptop through a parallel, serial, or USB port.

You may find a great bargain on a used bench oscilloscope on eBay — if you're willing to take a risk. We've seen some in the $100–200 price range.

Observing the ups and downs of voltage

Oscilloscopes give you a visual representation of an electrical signal. The vertical axis indicates the amount of voltage (also called *amplitude*), and the horizontal axis represents time. (Remember graphing equations in math class? Well, the display on a scope is really such a graph.) Oscilloscopes always sweep left to right, so you read the timeline of the signal from left to right, just as you'd read a line of English on a page.

The signal that you observe on the oscilloscope is a *waveform.* Some waveforms are simple, some are complex. (We introduce the concept of waveforms in Chapter 2.) Figure 13-4 shows the four most common waveforms that you encounter in electronics:

✔ **DC (direct current) waveform:** A flat, straight line, like the one you see here.

✔ **AC (alternating current) waveform:** This waveform undulates over time. The most common AC waveform is a sine wave (see Chapter 2 for more about sine waves), but you may also encounter triangle waves, sawtooth waves, and other AC waveform shapes.

✔ **Digital waveform:** A DC signal that alternates between 0 V (low, which indicates logical 0) and some pre-determined voltage (high, which indicates logical 1). The digital circuitry interprets the timing and spacing of the low and high marks. (You can read more about digital signals in Chapter 7.)

✔ **Pulse waveform:** This waveform shows a sudden change between a signal's low and high states. Most pulse waveforms are digital and usually serve as a timing mark, like the starter's gun at a race. The 555 timer IC we describe in Chapter 7 can be configured to operate as a *one-shot,* producing a single pulse that triggers other parts of the circuit to generate even more signals.

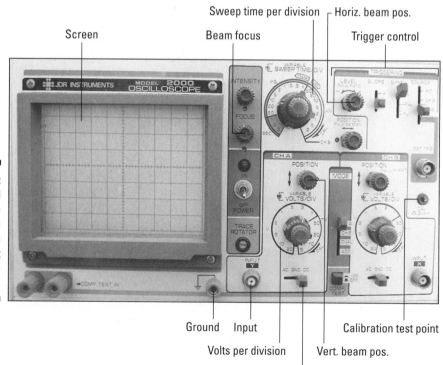

Figure 13-3:
A typical bench oscilloscope with its important controls identified.

Screen · Beam focus · Sweep time per division · Horiz. beam pos. · Trigger control

Ground · Input · Volts per division · Signal clamp · Vert. beam pos. · Calibration test point

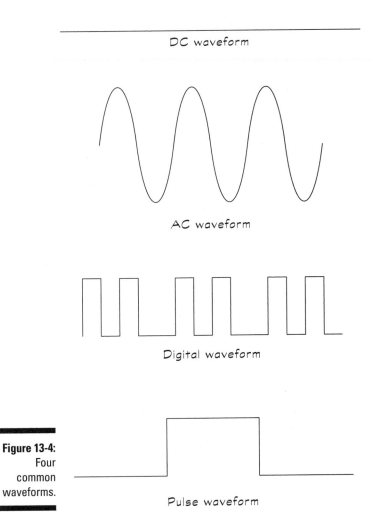

DC waveform

AC waveform

Digital waveform

Figure 13-4:
Four
common
waveforms.

Pulse waveform

Figure 13-5 shows what the AC waveform looks like on an oscilloscope screen. The display has a built-in grid to help you measure time along the X (horizontal) axis and voltage along the Y (vertical) axis. Using knobs on the front panel, you select the voltage scale (for instance, 5V/division) and *sweep time* (for instance, 10 ms/division, where ms is milliseconds, or thousandths of a second) of the display. As you adjust these settings, you see the voltage display change proportionally. You can read off a voltage level at a particular time by determining the position of the voltage on the grid and multiplying that by the voltage scale you've selected.

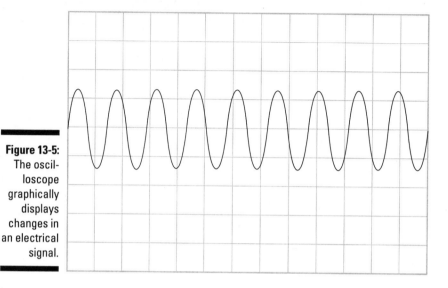

Figure 13-5:
The oscil-
loscope
graphically
displays
changes in
an electrical
signal.

A DC waveform's vertical position (amplitude) gives you the DC voltage reading. For AC signals, the oscilloscope display enables you to determine voltage levels as well as frequency (the number of cycles per second). If you count the number of horizontal divisions that one complete cycle occupies on the screen, and multiply that by the time scale (for instance, 10 ms/division), you get the *period, T,* of the signal (the time it takes for one cycle to complete). The frequency, f, is the reciprocal of the period; the formula for f looks like this: $f = 1/T$. (For more details, flip to the section "Testing the frequency of an AC circuit," later in this chapter).

Understanding oscilloscope bandwidth and resolution

One of the most important specifications for oscilloscopes is bandwidth. *Bandwidth,* measured in megahertz (MHz), is the highest frequency signal that you can reliably test with your oscilloscope. The average bandwidth of a low-cost bench scope falls in the 20–35 MHz range, which is fine for all but the most demanding applications. Specialized troubleshooting and repair, such as work on computers and ultra-high-frequency radio gear, may require bandwidths exceeding 100 MHz — which you can find among the pricier oscilloscopes. PC-based scope probes tend to have the lowest bandwidth, usually about 5–10 MHz. This bandwidth does the job for many tasks, from working with hobby circuits to servicing audio equipment.

Some enhanced features you should know about

Oscilloscopes have improved greatly over the years, with many added features and capabilities. Although you don't absolutely need any of the following features for routine testing, you may find them handy as you gain experience. Among the most useful features are

✔ **Delayed sweep:** When analyzing a small portion of a long, complex signal, this feature helps because you can zoom in on a portion of the signal and examine it. This is ideal when you work with television signals.

✔ **Digital storage:** This feature records signals in computerized memory for later recall. After you have it in the memory, you can expand the signal and analyze specific portions; again, helpful in television work. Digital storage also lets you compare signals, even if you take the measurements at different times.

As you may expect, these features can add to the cost of the scope. Balance the extra cost against the usefulness of these features.

Most scopes have a *resolution,* or accuracy, of 0.5 µs (that's microseconds, or millionths of a second) or faster. You can adjust the sweep time so that you can test signal events that occur over a longer time period, usually as long as a half a second to a second. Note that the screen can display signal events faster than 0.5 µs, but such a small signal may appear as a fleeting glitch or voltage spike.

The sensitivity of an oscilloscope indicates the Y-axis voltage per division. The low-voltage sensitivity of most average-priced scopes ranges from about 5 mV (that's millivolts, or thousandths of a volt) to 5 V. You turn a dial to set the sensitivity that you want. When you set the dial to 5 mV, each division on the scope grid represents a difference of 5 mV. Voltage levels lower than 5 mV may appear, but you can't accurately measure them. Most scopes show very low voltage levels (microvolt range) as a slight ripple.

Knowing When to Use an Oscilloscope

When you're testing voltage levels, you can often use multimeters and oscilloscopes interchangeably. The choice of which tool you use is yours, though for routine testing procedures, you may find the multimeter a little easier. In general, you may opt to use an oscilloscope for

✔ **Determining visually whether an AC or digital signal has the proper timing.** For example, you often need this test when you troubleshoot radio and television equipment. The service manuals and schematics for these devices often show the expected oscilloscope waveform at various points in the circuit so that you can compare. Very handy!

✔ **Testing pulsating signals that change too rapidly for a logic probe to detect.** Generally these are signals that change faster than about five million times a second (5 MHz).

✔ **Visually testing the relationship between two signals,** when using a *dual-trace oscilloscope,* a scope with two input channels. You may need to do this test when you work with some digital circuits, for example. Often one signal triggers the circuit to generate another signal. Being able to see both signals together helps you determine whether the circuit is working as it should.

✔ **Testing voltages,** if the scope is handy; but you can use your multimeter for testing voltages, too.

Rather than whipping out your oscilloscope for every test, you're better off using a multimeter for the following:

✔ Testing the resistance of a circuit

✔ Determining whether a wire or other part of a circuit has short-circuited (0 Ω resistance) or is open (infinite Ω resistance)

✔ Measuring current

✔ Testing voltages and various components, such as capacitors and transistors

Getting Your Oscilloscope to Work

An oscilloscope is a fairly complicated piece of equipment. To thoroughly understand its proper use, read the instruction manual that comes with the scope — or a book dedicated to the subject. This section gives you just a quick overview to get you started. You may also want to visit one of the electronics sites mentioned in the Appendix; many sites provide tutorials on using an oscilloscope.

Basic setup and initial testing

Before you use your oscilloscope for any actual testing, set its controls to a normal or neutral setting. Then you calibrate the scope, using its built-in test point, to make sure it's working correctly.

Refer to Figure 13-3, earlier in this chapter, to identify the various knobs and buttons on your scope as you go through these steps. Remember that your oscilloscope may look a bit different, and its knobs and controls may have slightly different names. Here are the steps for setting up your scope:

1. **Turn the scope on.**

   If it's the CRT bench-top variety, allow time for the tube to warm up. You may or may not see a dot or line on the screen.

2. **Set the Sweep Time/Division knob to 1 ms.**

   This setting is a good middle value for initial calibration.

3. **Set the Volts/Division knob to 0.5 V.**

   This setting is also a good middle value to use for initial calibration when you're testing low-voltage DC circuits.

4. **Set the Trigger Level control to Automatic (or midway, if it doesn't have an Automatic setting). Select AC Sync and Internal Sweep.**

5. **Select the Auto setting for both Horizontal Position and Vertical Position.**

   If your scope doesn't have an Auto setting, crank the knobs to their midpoints.

6. **Connect a test probe to the input.**

   If your scope has multiple input channels, use Channel A.

7. **Select Gnd (Ground) for the Signal Clamp, if your scope has this control.**

   On some scopes, this control may be called Signal Coupling.

8. **Connect the ground clip of the test probe to the designated ground connection on the scope (see Figure 13-6).**

   If your oscilloscope doesn't have a designated ground connection, clip the lead to any exposed metal point, such as the head of a screw.

9. **If your scope has a Signal Clamp switch, attach the center of the test probe to the calibration test point.**

   If your scope lacks a Signal Clamp switch, attach the center of the test probe to the ground point.

10. **Adjust the Vertical Position knob until the beam sits on the first division on the screen (see Figure 13-7).**

11. **Adjust the Horizontal Position knob until the beam is more or less centered on the screen.**

    You don't need to worry about making this setting exact.

12. **If your scope has a Signal Clamp switch, set it to DC. If you don't have a Signal Clamp switch, move the test probe from its ground connection to the calibration test point.**

Figure 13-6:
Connect the
ground of
the probe to
the ground
connec-
tion on the
scope.

Figure 13-7:
Adjust the
Vertical
Position
knob so that
the beam
sits on the
bottom of
the grid.

Many oscilloscopes use a test signal that appears as a relatively low-frequency square wave — a digital waveform that varies uniformly between low and high voltages. Consult the manual that came with your scope to see what voltage and frequency your scope produces with its built-in test calibration circuit.

For example, say that the signal should be 0.5 V peak-to-peak (indicated as 0.5 V p-p) at 1,000 Hz. Because you set the Volts/Division knob to 0.5 V and the test signal has amplitude of 0.5 V, the waveform spans one division on the screen.

By decreasing the Volts/Division setting, you can make the waveform larger. Do this adjustment when you need more accuracy. For example, if you set the Volts/Division knob to 0.1 V, a 0.5 V test signal spans five divisions.

Displaying and measuring signals

After you've set up and tested your oscilloscope, you're ready to use it to visualize what's going on in your circuits.

Do not use an oscilloscope to test AC voltage coming from a wall outlet unless you first take special precautions. The manual that came with your scope should outline those precautions. This book assumes you're using your oscilloscope to test *only* low-voltage DC circuits and low-voltage AC signals, such as those from a microphone. If you connect your scope directly to 120 VAC from a wall outlet, you can injure both you and your scope!

Here's a quick rundown of the steps that you perform to measure voltage signals with an oscilloscope:

1. **Attach a test probe to the scope input.**

   Note: Some scopes have several inputs, called *channels;* these steps assume you're dealing with just one input for now.

2. **Adjust the Volts/Division control to set the amplitude or voltage range.**

   For example, if the voltage you're testing is somewhere between 0 V and 5 V, use the 1 V per division range. With that setting, each volt corresponds to one division on the screen of the scope.

3. **Adjust the Sweep Time/Division control to set the time slice of the signal.**

   The *time slice* is the duration of the part of the signal that's shown on the scope. A shorter time slice shows only a brief portion of the signal, whereas a longer time slice shows you more of it. You can experiment with the Sweep Time/Division control until the signal looks the way that you want it to.

   If you're testing a low-voltage AC or pulsing digital signal, set this control so you can adequately see each cycle of the signal. If you're testing a steady DC signal, this control is less important because the signal doesn't change (much) over time. Choose a medium-range setting, such as 1 ms per division, to ensure consistent readings.

4. **Select the signal type (AC or DC) and the input channel.**

   Many scopes (such as the one shown in Figure 13-3) offer two input channels, labeled Channel A and Channel B, so you can measure and compare two signals simultaneously. If yours is a single-channel oscilloscope, it won't have an input-channel selector.

5. **Adjust the Vertical Position knob to set the 0 V reference level on the scope.**

   If you expect to view both positive and negative voltages, you should set the vertical position to the middle of the screen (at the fifth vertical division); if you expect to view just positive voltages, you should set the vertical position to the bottom of the screen (at the first vertical division). The vertical position you set in this step — with no input applied to the scope — specifies the 0 V reference level on the scope screen.

6. **If your scope has a trigger switch (as most do), set it to Auto.**

7. **When you've set up the oscilloscope properly, connect the test probe to the signal that you want to test.**

   Connect the ground of the probe to the ground of the circuit. Connect the probe itself to the circuit point that you want to test (you can see this setup in Figure 13-8).

8. **Read the waveform displayed on the screen.**

   Unless your scope has a direct readout function that displays numeric voltages on the screen, you need to correlate what you're seeing with the settings of the oscilloscope. To read a voltage level, you determine the vertical position of a point on the waveform display with respect to the 0 V reference position you set in step 5, and then multiply it by the voltage scale (Volts/Division) you've selected.

   Say, for instance, you've set the 0 V reference position to the middle of the screen (fifth vertical division), and the voltage scale to 2 V per division. A point on the displayed waveform that is 3.2 divisions above the middle of the screen represents a voltage of 6.4 V (3.2 division × 2 V per division). A point that is 1.5 divisions *below* the middle of the screen represents a voltage of –3.0 V (–1.5 divisions × 2 V per division).

If you have a dual-trace oscilloscope, you can display two signals at once. Dual-trace oscilloscopes have two sets of leads: one set for Channel A and one set for Channel B. You can move each trace up and down independently, so you can display one signal above the other or superimpose them. Say, for instance, you want to check the gain of an amplifier circuit. You connect the Channel A leads to the input to the amplifier, and the Channel B leads to the output of the amplifier. By lining up the traces so they're on top of each other, you can easily compare the signals and calculate the gain. This capability can be very helpful in analyzing your circuits.

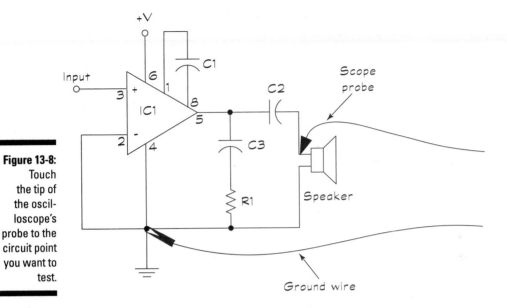

Figure 13-8:
Touch
the tip of
the oscil-
loscope's
probe to the
circuit point
you want to
test.

Testing, Testing, 1-2-3!

So if you've been reading along in this chapter, you now know a little bit
about what an oscilloscope is for, what it does, and how to use it to display
a signal. The following sections show you how to do a couple of quick tests.
These tests demonstrate how you use a scope for a variety of simple chores.
After you work through these tests, you're well on your way to becoming a
master scope user.

Does your battery have any juice?

A rudimentary test that will help familiarize you with the oscilloscope is measuring battery voltage. A battery produces a steady DC voltage, so the sweep
setting on the scope is irrelevant in this test. You just want to know what voltage the scope displays on the screen.

For this demonstration, dig around in your drawer and pull out a 9 V battery.
After running the procedure outlined in the section "Basic setup and initial
testing," earlier in this chapter, follow these steps to test your battery:

1. **Set the Volts/Division knob to 2 V.**

2. **Make sure the Vertical Position is set to the bottom of the screen (first
   vertical division).**

   This sets the 0 V reference level to the bottom of the screen.

3. **Attach the ground clip of the test probe to the negative terminal of the battery.**

4. **Attach the center of the test probe to the positive terminal of the battery.**

   If the battery is fully juiced, the line on the screen should fall approximately midway between the fourth and fifth vertical divisions. Because you have the Volts/Division knob set to 2 V, this line placement indicates that your battery has 9 V (4.5 divisions × 2 V per division). If the line on the screen falls well below the fourth vertical division, it may be time to retire that battery.

Dissecting your radio to display an audio waveform

Oscilloscopes can visually represent AC waveforms, such as the electrical signal that drives a speaker. This audio signal is complex because it's made up of constantly changing frequencies. These changing frequencies are what you hear as singing, talking, or the sound of musical instruments.

For this test, pry off the back of an ordinary battery-powered radio so that you can reach the two terminals on the speaker. After running the procedure outlined in the section "Basic setup and initial testing," earlier in this chapter, follow these steps to observe the signal driving your radio speaker:

1. **Adjust the Vertical Position knob until the beam sits in the middle of the screen (the fifth vertical division).**

   This way, you'll see voltage swings above and below 0 V.

2. **Set the Volts/Division knob to 1 V.**

3. **Set the Sweep Time/Division knob to 0.1 milliseconds (0.1 ms).**

4. **Attach the ground clip of the test probe to one of the speaker terminals.**

5. **Attach the center of the test probe to the other speaker terminal.**

6. **Turn on the radio and watch the display.**

7. **If at first you don't get much of a reading, try, try again by decreasing the Volts/Division setting.**

Here are some things to watch for when you perform this test:

✔ **The amplitude of the waveform increases and decreases as you change the volume on the radio.** This change happens because the volume control alters the signal voltage that you apply to the speaker.

> ✔ **By turning the Sweep Time Per Division knob, you can see finer details of the signal.**

If you have access to a *function generator* that can produce a variety of signals (as we describe in Chapter 16), you can use this same technique to take a look at its waveforms. Rather than a mish-mash of squiggly lines, you see a distinct sine wave, and observe how the display changes when you vary its frequency. You can also observe other types of waveforms, such as square waves and triangle waves.

Testing the frequency of an AC circuit

You can determine the frequency of an AC signal, and display the 60 Hz (50 Hz in some parts of the world) alternating current coming out of a wall outlet. Note however, the following caution before you plug anything into anything:

Although it's technically *possible* to plug the test probe of an oscilloscope directly into a wall socket — DON'T! Even. Think. About. It. Doing so poses a significant safety hazard.

Instead, you can test the frequency of your household current indirectly — and safely — using a phototransistor.

Yep, for this test, you need a *phototransistor* (a light-dependent transistor, detailed in Chapter 8) and a 10kΩ resistor. (For more about resistors, see Chapter 3.) Connect the phototransistor and resistor to a 9 V battery — like the setup in Figure 13-9. Then grab a household lamp outfitted with an incandescent bulb, and you're ready to go!

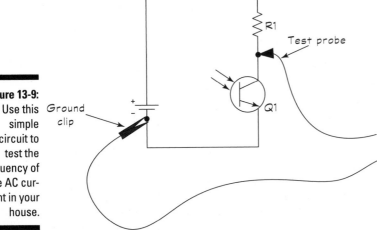

Figure 13-9: Use this simple circuit to test the frequency of the AC current in your house.

After following the setup-and-testing procedure outlined in the section "Basic setup and initial testing," earlier in this chapter, perform your test by following these steps:

1. **Adjust the Vertical Position knob until the beam sits in the middle of the screen (the fifth vertical division).**

2. **Set the Volts/Division knob to 1 V.**

3. **Set the Sweep Time/Division knob to 10 ms.**

4. **Attach the ground clip of the test probe to the negative terminal of the battery.**

5. **Attach the center of the test probe to the point where you have connected the phototransistor and resistor.**

6. **Turn on the light and note the ripple in the waveform.**

    For best results, don't shine the lamp directly into the phototransistor. Instead, direct the light away until you can see a sine wave. If your lamp is too close to the phototransistor, the transistor may become saturated (which is like an overload), and you won't see any signal change. Adjust the Volts/Division knob until you get a decent reading.

The phototransistor can detect very quick changes in light, which you see as the ups and downs of the signal on the scope screen, reacting much faster than your eyes. Try as you may, you won't be able to directly observe the lamp getting brighter or dimmer as the current increases and decreases; that's because of *persistence of vision* (your brain's insistence on seeing continuity of motion). But through the magic of electronics, you will "see" the light pulsing!

You must use an incandescent bulb for this test to work. Modern compact fluorescent lamps (CFLs) include electronic circuits designed to stabilize the current flowing through the gas-filled tube as the AC voltage alternates up and down.

The ripple on the scope screen represents the AC current pulsing through the incandescent lamp. But a phototransistor registers a flash of light each time the AC current goes positive or negative, so the ripple that you see changes *twice as often* as the household current flowing through the lamp. So the frequency of the waveform on the scope screen is twice the frequency of household current — expect it to be about 120 Hz.

Forget for the moment that you already know what the frequency of the displayed waveform is. To determine the frequency by reading the scope, you first calculate the period of the waveform — the time it takes to complete one up-and-down cycle — and then convert it to frequency by taking the reciprocal. To do this, you measure the distance from crest-to-crest of the waveform; in this case, it should be about 0.8 of a division. You then multiply that distance by the Sweep Time/Division that you set up (10 ms or 0.01 seconds) to get the period, like this:

Period (in seconds) = 0.8 division $\times$ 0.01 seconds/division = 0.008 seconds

Then you take the reciprocal of the period to get the frequency:

Frequency (in Hertz) $= 1/_{Period}$ (in seconds)

$= 1/_{0.008}$

= 125 Hz

Recalling that the frequency of the waveform on the scope screen is twice the frequency of the household current pulsing through the lamp, you divide 125 Hz by two to get your estimate of the frequency of household current: 61.5 Hz. That's pretty close to 60 Hz.

An oscilloscope enables you to calculate an *approximation* of the frequency of a signal. If you need something more accurate, you need a piece of test equipment called a *frequency counter.* These devices use a digital readout to display signal frequency, and they're accurate to within one in several hundred thousand hertz. Check out Chapter 16 for more about frequency counters.

Part III
Putting Theory into Practice

The 5th Wave   By Rich Tennant

"I think I've fixed the intercom. Just remember to speak into the ceiling fan when the doorbell rings."

In this part . . .

*U*nderstanding how electronics components control electric current is useful, but seeing this control in action is truly electronic nirvana. In this part, you find out how to set up simple circuits to demonstrate the principles of electronics. We show you how to use your multimeter — and a wee bit o' math — to gather enough evidence to prove that electronic components really do live up to their promises. You may even make a few LEDs glow, to show where and when electric current is flowing through your test circuits.

We also share with you the plans for several fun electronics projects you can build in less than 30 minutes — without breaking open your piggy bank. Using sensors, lights, and buzzers, you can put electronics to work to create a variety of practical — and, sometimes, entertaining — circuits. You may find that these projects spark some creative ideas for other circuits you can build to impress your friends and family.

Chapter 14

Exploring Some Learning Circuits

. . . .

In This Chapter

▶ Enforcing Ohm's Law with simple resistive circuits

▶ Witnessing the ups and downs of capacitor voltage

▶ Seeing the light with LEDs

▶ Zeroing in on a Zener diode

▶ Switching and amplifying current with transistors

▶ Thinking logically with NAND gates

. . . .

By now, you may be wondering if all that theoretical mumbo-jumbo discussed in Chapters 1–6 will really work in your very own home-grown electronics lab. Will your resistors really resist? When you charge up a capacitor, will it faithfully hang on to its electrical energy until you give it the go-ahead to discharge? Will your semiconductors actually conduct when you want them to? And are logic gates truly as logical as advertised?

If you're interested in seeing the principles of electronics in action, then you've come to the right chapter.

In this chapter, we show you simple circuits that quell electronic anxieties by demonstrating how various components and the laws that govern them work. You can build any of these circuits in less than ten minutes, and witness firsthand how one or more components shape electrical current. Using your multimeter as your guide — with some visual cues from a few strategically placed LEDs — you can get a real sense of what's going on deep down inside each circuit.

Getting Ready to Explore

To prepare to build the circuits in this chapter, make sure you have a few supplies at the ready. At a minimum, you need the following items to build and explore any (or all) of the circuits discussed in this chapter:

✔ **Solderless breadboard.** You should have at least one solderless bread-board to enable you to build, update, tear down, and re-build the circuits we discuss. If necessary, review the breadboarding techniques discussed in Chapter 11 so you know which holes are connected to which other holes before you start poking components in randomly.

✔ **Batteries.** All of the circuits in this chapter run off one or two 9 V batteries. It pays to have a few extra around in case you leave a circuit powered up for a long time. To make circuit building easier, buy two 9 V battery clips and solder 22-gauge solid wires onto the clip leads so you can easily connect and disconnect your batteries from any circuit, eliminating the need for a switch. (For more about soldering leads, see Chapter 11.)

✔ **Multimeter.** If you're reading this chapter, it's a pretty good bet you want to explore some circuits. To do that, you need a multimeter to measure voltage, resistance, and current as you observe what's happening to the electrons your circuits are manipulating. Review Chapter 12 if you need a refresher on how to wield a multimeter.

✔ **Calculator.** If you don't trust your brain to perform mathematical calculations accurately, bring your calculator to your workbench.

Carefully check your units of measurement. Convert values to ohms, amps, and volts before you calculate anything.

Just to make your shopping trip a little easier, here's a list of all the electronic components used in the learning circuits in this chapter:

✔ **Resistors:** With one exception, which is noted, resistors rated at $1/4$ W or $1/8$ W with 10 percent or 20 percent tolerance are fine. You'll need the following fixed resistors: one 220 Ω, one 330 Ω, two 470 Ω $1/2$ W, two 1 kΩ, two 10 kΩ, and one 100 kΩ.

✔ **Potentiometers:** One each: 10 kΩ, 100 kΩ, and 1 MΩ. To prepare each potentiometer for use in your solderless breadboard, solder 3-inch wires to each of its three leads. Remember that the middle lead is connected to the wiper arm inside the device, and the leads on either end are fixed.

✔ **Capacitor:** One 470 µF electrolytic capacitor rated at a minimum of 25 V.

✔ **Transistors:** Two general-purpose NPN bipolar transistors, such as 2N3904 or BC548.

✔ **LEDs:** Two standard 5mm LEDs of any color (same or different).

✔ **Diodes:** One 1N4731 4.3V 1W Zener diode.

✔ **IC:** One 4011 quad 2-input NAND gate IC.

For the first circuit we discuss, we walk you through the process of building it on the solderless breadboard. For subsequent circuits, we simply show you the schematic diagram and leave the construction up to you. As you build each circuit, use a row along the top of your solderless breadboard for connections to the positive side of the power supply, and a row along the bottom for ground connections (the negative side of the power supply), as we explain and illustrate for the first learning circuit.

Seeing Is Believing: Ohm's Law Really Works!

Chapter 3 looks at one of the most important principles in electronics: Ohm's Law: The voltage, V, dropped across a component with a fixed resistance, R, is equal to the product of the current, I, flowing through the component and the value of the resistance (R). Ohm's Law governs all resistive electronic components. You can sum it up with this equation:

$$V = I \times R$$

You can also rearrange the terms of that equation to get two other equations which really say the same thing in two other ways:

$$I = \frac{V}{R}$$

and

$$R = \frac{V}{I}$$

You use Ohm's Law to help you analyze circuits, from simple series circuit to more complex series-parallel circuits. In this section, you can test Ohm's Law to make sure it works (it does!) and to take your first steps in circuit analysis.

As you explore these circuits, keep in mind another important rule (Kirchoff's Voltage Law) that applies to every circuit: *The voltage rises and drops around a circuit always sum to zero.* In DC-powered circuits, such as the ones explored in this section, another way to say the same thing is that the voltage drops that happen across *all* the components in a circuit add up to the source voltage. (If you'd like to review these concepts, refer to Chapter 2.)

Analyzing a series circuit

Figure 14-1 shows a simple series circuit containing a 9 V battery, a 470 Ω resistor (R1), and a 10kΩ potentiometer, or variable resistor (R2). We've also labeled the voltage drops across R1 and R2. For this circuit, you will connect the middle (wiper) lead and one outer (fixed) lead of the potentiometer together — as is common practice when using a potentiometer (pot) as a two-terminal variable resistor. This way, R2 is the resistance from the combined leads to the other outer (fixed) lead. By dialing the pot, you can vary the resistance of R2 between 0 (zero) Ω and 10kΩ.

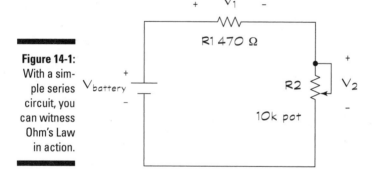

Figure 14-1:
With a simple series circuit, you can witness Ohm's Law in action.

Before using a pot in your circuit, use your multimeter — set on Ohms — to measure the resistance of the pot (from the wiper lead to the fixed lead that is not connected to the wiper). Turn the pot all the way in one direction so your multimeter reads 0 (zero) Ω.

Building a simple series circuit

Refer to Figure 14-2 for making connections on your solderless breadboard. Build the circuit one step at a time, connecting the battery last, as follows:

1. **Insert the fixed resistor (R1).** Plug one lead (either one) into a hole along the top row of the breadboard and the other lead into a hole in the center section of the breadboard.

2. **Insert the potentiometer (R2).** Plug the fixed lead (the one that's not connected to the wiper lead) into a hole in the center section, in the same column as the lead for resistor R1, and plug the other two leads into holes along the bottom row.

Figure 14-2:
Setting up
the simple
series cir-
cuit on a
solderless
breadboard
is a snap.

3. **Insert the battery.** Plug the lead for the negative terminal into a hole along the bottom row, and then plug the lead for the positive terminal into a hole along the top row to complete the circuit.

Now it's time to probe around your live circuit using your multimeter.

Adding up voltages around a circuit

First you can explore some voltages. Set your multimeter to DC Volts with a range of 10 V, and follow these steps:

1. **Start with a simple test to determine the exact voltage of your battery: Touch the black voltmeter probe to the negative terminal of the battery and the red voltmeter probe to the positive terminal of the battery.**

   You can do so by gently inserting the tip of each probe into a hole in the breadboard that is connected to the point you're measuring in the circuit.

   You should see a reading of roughly 9 volts (for a fresh battery). Note the reading.

2. **Test the voltage across the pot (R2). To do so, place the black voltmeter probe at the point where the R2 is connected to the negative terminal of the battery, and place the red probe at the point where R1 and R2 are connected.**

What reading do you get? What reading should you expect to get, and why? (Hint: The pot is turned all the way down to zero Ω.)

3. **Test the voltage across R1. Take care to orient the voltmeter probes so that you get a positive voltage reading, placing the black probe at the point where R1 meets R2, and the red probe at the point where R1 is connected to the battery.**

   What reading do you get, and why? If you add the voltages you've measured together, you should get the full battery voltage, as follows:

   $$V_1 + V_2 = V_{battery}$$

   Try adjusting the pot to a few different settings and measuring V_1 and V_2 again. Do they always add up to $V_{battery}$?

Zeroing in on Ohm's Law

To test Ohm's Law, zero the pot again — and think about what current you expect to see in this series circuit with the pot turned all the way down to zero ohms. Then, with the probes disconnected from the circuit, change the multimeter setting to DC Amps, and set the range to 200 mA. (You may want to dial this down to a 20 mA range, but it's better to start with a larger range and dial it down rather than start small and go large.)

Remember that to measure current, you need to insert your meter in series with whatever you're measuring current through. But before you break the circuit, disconnect the battery by simply removing one of its leads from the breadboard. It's good to get into the habit of removing power before fiddling with circuit components.

To test Ohm's Law, follow these steps:

1. **With the power disconnected, move the R2 (pot) lead that is connected to R1 to another column in your breadboard.**

   Now you've broken the circuit.

2. **Connect the red lead of your meter to the open side of R1 and the black lead of your meter to the open side of R2. Then connect the battery to the circuit.**

   You should get a current reading. If the reading is less than 20 mA, you can dial down your meter setting to the 20 mA range to get a more precise reading. Is the current reading what you expect it to be? Are you convinced that Ohm's Law applies to your circuit?

3. **With your meter still connected between R1 and R2, slowly dial up the potentiometer and observe the change in current.**

   Is the reading heading in the right direction?

4. **Dial the pot all the way to the end.**

   Is the current reading what you expect it to be? Remember that the current through a series circuit is limited by all of the resistances in the circuit, and that resistances in series add up. (Keep in mind, too, that the meter will add a certain amount of resistance, albeit tiny, to the circuit.)

5. **Dial the pot down until the current reading is about 10 mA and calculate the value of the resistance R2 that you expect.**

6. **Disconnect the battery and then remove the pot from the circuit.**

7. **Switch your multimeter setting to Ohms with a range of 2kΩ. Then measure the resistance across the pot leads.**

   Is the resistance about what you expected?

You can experiment as much as you want, varying the pot and measuring currents and voltages to verify that Ohm's Law really does work. You'll also see that the voltage drops across R1 and R2 vary as the pot resistance changes — and that those voltages always sum to the battery voltage. Try filling in the chart given here as you run your tests; add rows for other pot settings.

R2	Current	V_2	V_1
0 Ω			
	10 mA		
		5 V	
10 kΩ			

Dividing up voltage

Using the same series circuit (shown in Figure 14-1), you can test the concept of the voltage divider discussed in Chapter 3. Say, for instance, you need to supply a circuit with 5 V, but you've got a 9 V power supply. You can divide the battery voltage between two or more resistive components, adjusting the resistances so that one of the voltages is 5 V.

Using Ohm's Law, you know that the voltage across R2 is equal to the current, I, through the circuit times the resistance, R2. You also know that the current, I, is equal to the battery voltage divided by the total resistance of the circuit, which is R1 + R2. From these two equations, you can write an expression for the voltage across R2:

$$V_2 = I \times R2$$

$$= \frac{V_{battery}}{R1 + R2} \times R2$$

$$= \frac{R2}{R1 + R2} \times V_{battery}$$

Notice that the expression on the right side of this equation is what you get when you multiply a ratio of resistance by the supply voltage. With the pot out of the circuit, dial it to some point and measure its resistance. Given this resistance, what would you expect the voltage across the pot to be? Insert the pot into the circuit, power it up, and measure the voltage across the pot. Is it close to your estimated value?

Use the following chart to test your voltage divider for a range of R2 values. What do you notice about the value of V_2 compared to the battery voltage when R2 is much larger than R1? Does this make sense to you?

R2	Estimated $V_2 = \dfrac{R2}{R1 + R2} \times V_{battery}$	Measured V_2
50 Ω		
100 Ω		
470 Ω		
1 kΩ		
5 kΩ		
8 kΩ		
10 kΩ		

Now, suppose you want to design a voltage-divider circuit to produce 5 V across R2. To do so, you calculate the specific value of R2 that will result in 5 V across R2. That means plugging in 5 V for V_2 in the voltage-divider equation shown earlier; doing so determines the value of R2. (Note that you don't need to know the current passing through the circuit in order to design the voltage divider.) After a bit of mathematical manipulation (which we won't go into here because this isn't a math book), and assuming your battery measures 9 V, you end up with the following equation for R2:

$$R2 = {}^5\!/_4 \times R1 = 1.25 \times R1$$

Because R1 equals 470 Ω (nominally, anyway), you should expect to produce 5 volts across R2 when the pot is set to about 588 Ω (that's 1.25×470 Ω).

Here's how to test this calculation:

1. **Set your multimeter to DC Volts with a range of 10 V and measure the voltage across the pot.**

2. **Dial the pot until the voltage reading is about 5 V.**

3. **Remove the pot from the circuit and measure the resistance.**

   The resistance value should be in the vicinity of 588 Ω.

You can test the voltage divider concept on your own. Pick a value of R2 and adjust the pot to that value. Calculate the expected voltage across R2, and then measure the actual voltage with the pot in the circuit. You get the idea.

Parallel parking resistors

Want to see current split right before your very eyes? Set up the circuit shown in Figure 14-3 and measure each current, following these steps:

1. **Set up your multimeter. Set it to DC Amps, with a range of 20 mA.**

2. **Measure the supply current, I_1, as follows:**

   a. Break the circuit between the battery and R1.

   b. Insert the multimeter in series with the battery and R1, with the red probe connecting to the positive battery terminal and the black probe connecting to the open side of the resistor.

   c. Note the current reading.

3. **Remove the multimeter, and then reconnect the battery and the resistor.**

4. **Measure the branch current, I_2, as follows:**

   a. Remove one of the leads of resistor R2.

   b. Insert the multimeter in series with R2, using the proper lead orientation.

   c. Note the current reading.

5. **Remove the multimeter, and then reconnect the R2 lead into the circuit.**

6. **Measure the other branch current, I_3, as follows:**

   a. Remove one of the leads of R3.

   b. Insert the multimeter in series with R3, using the proper lead orientation.

   c. Note the current reading.

7. **Remove the multimeter and turn it off. Then reconnect the R3 lead into the circuit.**

Unless the forces of nature have changed (or your multimeter has gone ker-phlooey), the branch currents you've measured should add up to the supply current you measured: $I_2 + I_3 = I_1$.

Figure 14-3:
The sup-
ply current
splits
between
the two
branches of
this circuit.

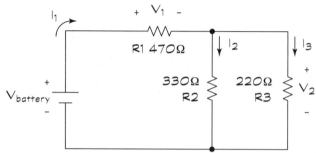

You can use math to calculate these currents by applying Ohm's Law and the rules for series and parallel resistances. To calculate the total supply current, I_1, you determine the total resistance of the circuit, R_{total}, and apply Ohm's Law using the battery voltage.

In Chapter 3, we discuss how to calculate the equivalent resistance, R_{total}, of a combo series and parallel circuit like the one in Figure 14-3, as follows:

$$R_{total} = R1 + \left(R2 \| R3\right)$$
$$= R1 + \frac{R2 \times R3}{R2 + R3}$$

If you run the numbers for the resistor values chosen here, you should find that $R_{total} = 602\ \Omega$.

Then you calculate the total supply current as follows:

$$I_1 = \frac{V_{battery}}{602\ \Omega}$$

Assuming $V_{battery} = 9$ V, you get

$$I_1 = \frac{9\ V}{602\ \Omega} \approx 0.015\ A = 15\ mA$$

Now that you know I_1, you can calculate the voltage V_1 across resistor R1, as follows:

$$V_1 = I_1 \times R1 = 0.015\ A \times 470\ \Omega \approx 7\ V$$

If 7 V is dropped across R1, then V_2, the voltage across the parallel resistors, must be 2 V (9 V – 7 V). Using V_2 and applying Ohm's Law to each parallel resistor, you can calculate the branch currents:

$$I_2 = V_2/R2 = 2\ V/330\ \Omega \approx 0.006\ A = 6\ mA$$
$$I_3 = V_2/R3 = 2\ V/220\ \Omega \approx 0.009\ A = 9\ mA$$

Amazingly, the two branch currents add up to the supply current.

Use your multimeter as a voltmeter, and verify your voltage calculations. Then, for even more fun, swap out the fixed resistor R3 and sub in a 10kΩ potentiometer. Vary the resistance and observe what happens to the currents and voltages. By the time you finish these tests, you'll be an Ohm's Law expert!

Charging and Discharging a Capacitor

In this experiment, you can observe a capacitor charging up, holding its charge, and discharging. You can also vary the time it takes to charge and discharge a capacitor. Take a peek at Chapter 4 to review how capacitors work and how you can control their operation.

Watching your charges go up and down

The circuit in Figure 14-4 is really two circuits in one. The changeover switch alternates between positions labeled "charge" and "discharge," creating two circuit options:

- ✔ **Charging circuit:** When the switch is in the charge position, the circuit consists of the battery, resistor R1, and the capacitor, C. Resistor R2 is not connected to the circuit.

- ✔ **Discharge circuit:** When the switch is in the discharge position, the capacitor is connected to resistor R2 in a complete circuit. The battery and R1 are disconnected from the circuit (they are open).

Use a jumper wire for the changeover switch. Poke one end of the jumper wire into your breadboard so it's electrically connected to the positive side of the capacitor. Then you can use the other end to connect the capacitor to R1 or R2. You can also leave the other end of the jumper wire unconnected, which we suggest you do later in this section. You'll see why.

Figure 14-4:
Move the
switch
in this
circuit to
alternately
charge and
discharge
the
capacitor.

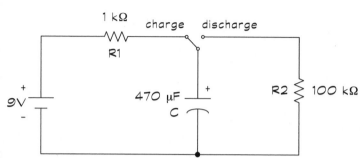

Figure 14-4:
Move the switch in this circuit to alternately charge and discharge the capacitor.

Set up the circuit using a 9 V battery, but don't connect the free end of the jumper-wire switch to anything yet. Make sure you properly orient the electrolytic capacitor by connecting the negative side of the cap to the negative terminal of the battery (or it may explode, and you don't want that).

To watch your capacitor charge up, hold its charge, and then discharge, follow these steps:

1. **Set your multimeter to DC Volts with a range of 10 V, and connect it across the capacitor (red lead connected to the positive side of the cap, black lead connected to the negative side of the cap).**

2. **Charge up the capacitor: Connect the changeover switch to the "charge" position (at R1) and observe the voltage reading on your meter.**

    You should see it rise to approximately 9 V — but not instantaneously — as the capacitor charges up through resistor R1. This should take a couple of seconds.

3. **Put the capacitor in a holding pattern: Remove the end of the jumper wire and just let it hang. Observe the reading on your voltmeter.**

    It should continue to read 9 V or thereabouts. (You may see the reading decrease a teeny bit; capacitors can and do leak charge.) The capacitor is holding its charge (really, holding electrical energy), even without the battery connected.

4. **Let the capacitor discharge: Connect the free end of the jumper wire to the "discharge" position (at R2) and observe the voltage reading on your meter.**

    You should see it decrease fairly slowly, as the capacitor discharges through resistor R2 to 0 V. This should take several minutes.

    Chapter 4 explains that a capacitor in a simple RC circuit reaches nearly its full charge at approximately five times the RC time constant, T. T is simply the value of the resistance (in ohms) times the value of the capacitance (in farads). So you can calculate the time it takes to charge and discharge the capacitor in your circuit as follows:

Charge time $= 5 \times R1 \times C$

$\qquad = 5 \times 1{,}000 \; \Omega \times 0.000470 \; F$

$\qquad = 2.35 \; s$

Discharge time $= 5 \times R2 \times C$

$\qquad = 5 \times 100{,}000 \; \Omega \times 0.000470 \; F$

$\qquad = 235 \; s \approx 3.9 \; min$

Is this what you observed? Repeat the charging and discharging experiment and see if your calculations seem about right.

Varying the RC time constant

By inserting a couple of potentiometers in your charging and discharging circuits, you can observe different RC time constants and watch your capacitor charge and discharge at different time intervals. Here's all you have to do:

1. **Set up the circuit in Figure 14-5, leaving the switch open (one end of the jumper wire disconnected from the board).**

   Notice that there is a 1kΩ resistor in series with the capacitor. That is to protect the capacitor by limiting current flow regardless of the potentiometer setting.

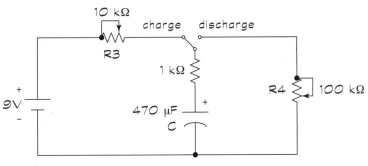

Figure 14-5:
Adjust the potentiometers in this circuit to vary the capacitor charging and discharging times.

The RC time constant of your charging circuit is determined by the total resistance in the charging circuit times the capacitance. The total resistance in the charging circuit is the sum of the fixed resistance (1kΩ) and the variable resistance dialed up on potentiometer R1. So you can calculate the capacitor charge time with this equation:

Charge time $= 5 \times (R3 + 1000) \times C$

If you set R3 to 0 Ω, your total resistance is simply 1,000 Ω, and you should expect the capacitor to charge up in approximately 2.35 seconds, just as it did in the preceding section ("Watching your charges go up and down").

2. **Move the jumper wire to the charge position and observe the reading on the voltmeter.**

The RC time constant of your discharge circuit is determined by the total resistance of the discharge circuit times the capacitance. The total resistance in the discharge circuit is the sum of the fixed resistance (1kΩ) and the variable resistance of potentiometer R4. So you can find the capacitor discharge time by using this equation:

Discharge time = 5 × (R4 + 1,000) × C

If you set R4 to its maximum resistance, 100kΩ, you should expect a discharge time of approximately 237 seconds or 4 minutes (nearly the same as in "Watching your charges go up and down").

3. **Switch the jumper wire to the discharge position and observe the voltmeter reading.**

4. **After the capacitor has discharged, remove the jumper wire.**

5. **Adjust R3 to its maximum value, 10 kΩ.**

6. **Move the switch to the charge position and observe the meter reading.**

Did you notice how much longer it takes to charge the capacitor? Did it seem to charge quickly at first, and then slow down a bit?

In Chapter 4, you can see the voltage versus time waveform of a charging/discharging capacitor. The waveform shows that at first, a capacitor charges quickly, and then it charges more slowly. The same holds true when a capacitor discharges: It initially discharges quickly, and then more slowly.

7. **Remove the jumper wire.**

8. **Adjust R4 to 0 Ω.**

9. **Move the switch to the discharge position and observe the meter reading.**

Did the capacitor discharge rather quickly? Expect it to discharge in about 2¹/₂ seconds.

Try adjusting each potentiometer to different values, and alternately charging and discharging the capacitor. If you really want to have fun, get out your kitchen timer and try to time the charge/discharge cycles. Then remove each potentiometer, measure each resistance, and calculate 5RC. Do your calculations roughly match your observations?

Dropping Voltages across Diodes

Diodes, which we discuss in Chapter 6, are like one-way valves for electrical current. By applying a small voltage from anode to cathode, current flows in the forward direction and the voltage drop across the diode remains fairly constant, even as the current increases.

In this section, you witness firsthand some of the ways diodes are used in electronic circuits. You can also use your multimeter to explore voltage drops and current in circuits with diodes.

Turning on an LED

For red, yellow, and green LEDs, a forward voltage of about 2.0 V turns on the valve, so to speak, allowing current to flow and the LED to light up. LEDs can carry currents of up to about 20 mA. (Check the ratings of the particular LED you choose.)

The circuit in Figure 14-6 is designed to demonstrate the on/off operation of an LED, and how increasing current strengthens the light emitted by the diode. Here's all you need to do:

1. **Dial a 10kΩ pot to its maximum resistance.**

2. **Set up the circuit shown in Figure 14-6, using a standard red, yellow, or green LED.**

   Make sure you orient the LED properly, with the cathode (negative side) connected to the negative battery terminal. On many LEDs, the shorter of the two leads is the cathode.

Figure 14-6: Use this circuit to turn an LED on and off, and to vary the intensity of the light.

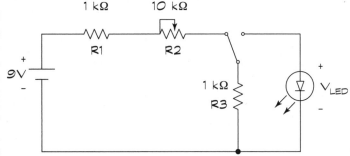

3. **Set your multimeter to DC Volts with a range of 10 V and place the leads across the LED.**

   Is the LED lit up? What voltage reading do you get? The LED voltage should be less than 1 V, which is not enough to turn the diode on.

4. **Dial the pot down slowly, keeping your eye on the LED. When the LED turns on, stop dialing the pot.**

   Observe the reading on the multimeter. Is the LED voltage close to 2.0 V?

5. **Continue to dial the pot down as you watch the LED.**

   What is happening to the light?

6. **Dial the pot all the way down to 0 Ω and observe the intensity of the LED. Note the voltage reading on your multimeter.**

   Did the LED voltage change much as the light intensity increased?

To understand why the LED was off when the pot was set to 10kΩ, and then turned on as you lowered the pot resistance, consider the circuit if you remove the LED. The circuit is a voltage divider, and the voltage across resistor R3 (which is the same as the LED voltage) is given by a ratio of resistance times the supply voltage:

$$
\begin{aligned}
V_{LED} &= \frac{R3}{R1 + R2 + R3} \times 9\ V \\
&= \frac{1,000}{1,000 + R2 + 1,000} \times 9\ V \\
&= \frac{1,000}{2,000 + R2} \times 9\ V
\end{aligned}
$$

If the resistance of the potentiometer is high (say, 10 kΩ), the voltage across the LED is rather low (about 0.75 V). When the resistance of the pot is low enough, the voltage across the LED rises enough to turn the LED on. V_{LED} will climb to about 2.0 V when R2 falls down to about 2.5 kΩ. (Plug in 2,500 for R2 in the equation above and see for yourself!) Of course, your particular LED may turn on at a slightly different voltage, say in the range of 1.7 V to 2.2 V. If you measure the resistance of your pot at the point at which your LED turns on, you may see a somewhat lower or higher resistance value than 2.5 kΩ.

You can also observe the current flowing through the LED by following these steps:

1. **Break the circuit between the cathode (negative side) of the LED and the negative battery terminal.**

2. **Insert your multimeter in series where you broke the circuit, and set it to measure DC Amps.**

3. **Start with the pot turned all the way up to 10 kΩ. As you dial the pot down, observe the current reading. Note the reading when the LED**

first turns on. Then continue to dial the pot down and observe the current readings.

You should see the current increase to over 5 mA as the light intensifies.

Overall, you should notice that the LED turns on when its voltage approaches 2.0 V, with just a tiny current passing through it at this point. As you increase the current through the LED, the light brightens, but the voltage across it remains fairly steady.

Clipping voltages

When a large enough reverse-bias voltage is applied to a Zener diode, it maintains a steady voltage drop, even as the current through it increases within a certain range. (Check out Chapter 6 for more detail on Zener diodes.) Follow these steps to test the voltage drop:

1. **Set up the simple voltage divider circuit in Figure 14-7, connecting two 9 V batteries in series to create the 18 V DC source.**

   Be sure to use resistors rated at ½ W for this circuit.

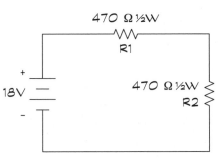

Figure 14-7: This simple voltage divider splits the supply voltage evenly between two resistors.

2. **Measure the total supply voltage, and then measure the voltage across R2.**

   It should be roughly half the supply voltage, or about 9 V.

3. **Measure the voltage across R1.**

   It should be about 9 V. The resistors are dividing the supply voltage equally.

4. **Now modify the circuit, as shown in Figure 14-8, by placing a 4.3V 1W Zener diode (1N4731) across R2, with the anode (positive side) of the diode connected to the negative battery terminal.**

5. **Measure the voltage across R2, which is the same as the voltage across the Zener diode.**

   Is it still about 9 V? Or is it roughly 4.3 V? (Zener diode tolerance — the variability of the component's actual voltage — can be ±10%, so the voltage can vary from 3.9 V to 4.7 V.)

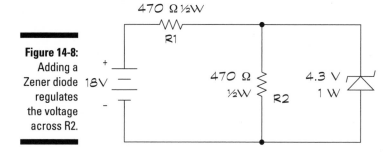

The Zener diode is regulating the voltage across R2. So where did the rest of the supply voltage go?

6. **Measure the voltage across resistor R1.**

   Has it increased since you added the Zener diode? (It should measure roughly 13.7 V.)

7. **Now remove the power and replace resistor R2 with a 10 kΩ potentiometer (see Figure 14-9).**

8. **Power up the circuit and measure the voltage across R2.**

9. **Vary the potentiometer from 0 Ω to 10 kΩ and observe the voltage reading.**

   Did you see the voltage rise steadily from 0 V until it reached the Zener voltage, and then remain at that voltage as you continued to increase resistance R2? The Zener diode holds the output voltage steady, even if the load on the circuit (represented by the potentiometer) varies.

470 Ω ½W

R1

18V

10 kΩ
R2

4.3 V
1 W

Gaining Experience with Transistors

In the next sections, you see how tiny transistors are used to control the current in one circuit (at the output of the transistor) using electronic components in another circuit (at the input of the transistor). That's what transistor action is all about!

Amplifying current

You can use the circuit in Figure 14-10 to demonstrate the amplification capabilities of a transistor. Here's all you need to do:

1. **Build the circuit using a general-purpose NPN bipolar transistor, such as a 2N3904 or BC548.**

   Be careful to connect the base, collector, and emitter leads properly.

2. **Dial the potentiometer all the way up so that the resistance is 1MΩ.**

   You'll probably see a tiny glow from LED2, but you may not see any light coming from LED1 — although there *is* a teeny current passing through LED1.

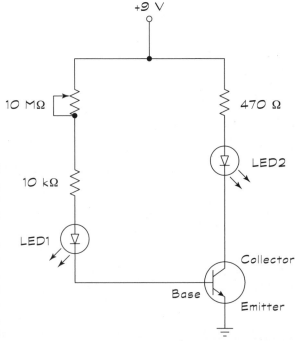

Figure 14-10: A pair of LEDs helps you visualize the amplification capabilities of a transistor.

Measuring teeny tiny currents

The base current of the bipolar transistor in Figure 14-10, which passes through LED1, is very small, especially when the pot is set to its maximum resistance. If you'd like to measure this teeny current, you can do it in these different ways:

✔ You can make the measurement directly, by breaking the circuit on one side of LED1, inserting your multimeter, and setting it to DC Amps. (The current is so small that it may not register on your meter.)

✔ You can measure the current indirectly, using Ohm's Law to help you. The same current that passes through LED1 and into the base of the transistor also passes through two resistors: the 10 kΩ resistor and the potentiometer. You can measure the voltage drop across either resistor and divide the voltage reading by the resistance. (Remember, Ohm's Law tells you that the current passing through a resistor is equal to the voltage across the resistor divided by the resistance.)

✔ If you really want an exact measurement, power the circuit down, pull the resistor out of the circuit, and measure its exact resistance with your multimeter. Then perform the current calculation. Using this method, we measured a base current of 6.1 μA (that's 0.0000061 A).

3. **Now slowly dial the pot down and observe the LEDs.**

You should see LED2 getting steadily brighter as you dial the pot down. At some point, you will start to see light from LED1 as well. As you continue to dial the pot down, both LEDs will glow brighter, but LED2 is clearly much brighter than LED1.

You're witnessing transistor action: The tiny base current passing through LED1 is amplified by the transistor, which allows a much larger current to flow through LED2. You see a dim glow from LED1 due to the tiny base current, and a bright glow from LED2 due to the stronger collector current. You can measure each current, if you'd like. (See the sidebar "Measuring teeny tiny currents" for a tip on how to measure the small base current.)

With the pot set to 1 MΩ, we measured a base current of 6.1μA (that's 0.0000061 A) and a collector current of 0.8 mA. Dividing the collector current by the base current, we find that the current gain of this transistor circuit is 131. With the pot set to 0Ω, we measured a base current of 0.61 mA and a collector current of 14 mA, for a current gain of approximately 23. Pretty intense!

The switch is on!

The circuit in Figure 14-11 is a touch switch. It uses a pair of NPN transistors to amplify a really teeny base current enough to light the LED. This piggyback configuration of two bipolar transistors, with their collectors connected and the emitter of one feeding into the base of the other, is known as a *Darlington pair.*

To test it, set up the circuit, using any general-purpose NPN transistors (such as 2N3904 or BC548). Close the circuit by placing your finger across the open circuit shown in the figure (don't worry, you won't get hurt). Did the LED turn on? When you close the circuit, your skin conducts a teeny tiny current (a few microamps), which is amplified by the pair of transistors, lighting the LED. (Touching, isn't it?)

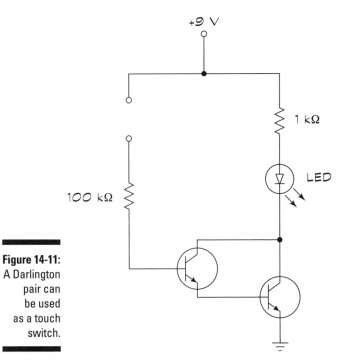

Figure 14-11:
A Darlington
pair can
be used
as a touch
switch.

Using Your Logic

Tiny digital circuits called *logic gates* accept one or more bits (binary digits) as inputs and produce an output bit that depends on the function of the particular gate. (You can read up on logic gates in Chapter 7.) Logic gates form the basic building blocks of advanced digital systems, such as the microprocessor that does all the thinking in your computer.

Inside each logic gate is a bunch of electronic components strung together in just the right way to perform the designated logic function. You find logic gates housed in integrated circuits (ICs) complete with several pins that allow you to access the inputs, outputs, and power connections of the digital circuit inside.

In this section, you find out how to make the right connections to a NAND logic gate IC, and you watch the output change as you fiddle around with various combinations of inputs. Then you find out how to create another type of logic gate, an OR gate, by combining NAND gates in just the right way.

Seeing the light at the end of the NAND gate

The circuit in Figure 14-12 uses an LED to indicate the high or low state of the output of a two-input NAND gate. Set up the circuit, using one of the four NAND gates on the quad 2-input NAND 4011 IC. (That's a CMOS chip that is sensitive to static, so be sure to review the precautions outlined in Chapter 9 to avoid damaging the chip.) For the switches, you can simply use jumper wires, inserting one end into your solderless breadboard, and moving the other end to close or open the switch.

Remember that the output of a NAND ("NOT AND") gate is high whenever either or both inputs are low, and the output of a NAND gate is low only when both inputs are high. (You can review the functionality of basic logic gates in Chapter 7.) "High" is defined by the positive power supply (9 V) and "low" is 0 V.

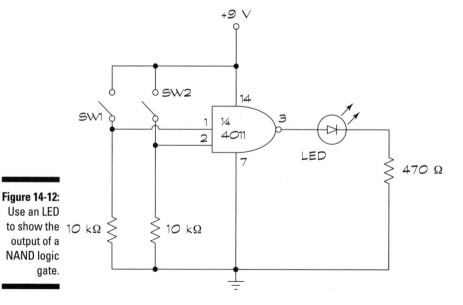

Figure 14-12:
Use an LED to show the output of a NAND logic gate.

When you close one of the switches, you make that input high because you connect the positive power supply voltage to the input. When you open one of the switches, you make that input low because it's connected through a resistor to ground (0 V).

Test the functionality of the NAND gate by trying all four combinations of open and closed switches, filling in the chart given here (which is essentially a truth table).

Input 1	Input 2	Output (High = LED on; Low = LED off)
Low (SW1open)	Low (SW2 open)	
Low (SW1 open)	High (SW2 closed)	
High (SW1 closed)	Low (SW2 open)	
High (SW1 closed)	High (SW2 closed)	

Did you see the LED light up when either or both switches were open? Did the LED turn off when both switches were closed? Be sure to tell the truth!

Turning three NAND gates into an OR gate

You can combine several NAND gates to create any other logical function. In the circuit in Figure 14-13, three NAND gates are combined to create an OR gate. The inputs to the OR gate are controlled by switches SW3 and SW4. The output of the OR gate is indicated by the on/off state of the LED.

Each of the two NAND gates on the left functions as a NOT gate (or inverter). Each NAND gate ties the inputs together so that a low input produces a high output, and a high input produces a low output. The NAND gate on the right produces a high output when either or both of its inputs are low, which happens when either or both switches (SW3 and SW4) are closed. The bottom line is that if either or both switches are closed, the output of the circuit is high. That's an OR gate!

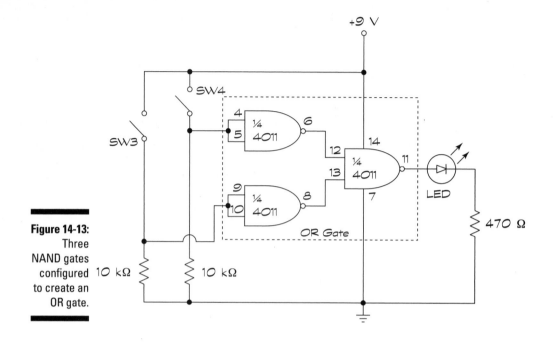

Figure 14-13:

Set up the circuit, being careful to avoid static. You can use the remaining three NAND gates on the 4011 IC that you used to build the circuit in Figure 14-12. Test the functionality by opening and closing the switches. The LED should turn on when either or both switches are closed.

Chapter 15

Great Projects You Can Build in 30 Minutes or Less

*G*etting up to speed on electronics really pays off when you get to the point where you can actually build a project or two. In this chapter, you get to play with several fun, entertaining, and educational electronics gadgets that you can build in half an hour or less. We selected the projects for their high cool factor and their simplicity. We've kept parts to a minimum, and the most expensive project costs under $15 or so to build.

We've given you some detailed procedures for the first project, so work through that one first. Then, you should be able to follow the circuit schematics and build the rest of the projects on your own. Check back to Chapter 10 if you need a refresher on schematics, and browse through Chapter 2 if you'd like to review basic circuit concepts. And if the projects don't seem to work as advertised (it happens to the best of us), review Chapter 12, arm yourself with a multimeter, and start troubleshooting!

Getting What You Need Right Off the Bat

You can build all the projects in this chapter, except for the electronic compass, on a solderless breadboard. Of course, feel free to build any of the projects on a regular soldered circuit board, if you want to keep them around. There's more detail about breadboarding and building circuits in Chapter 11. If you get stuck on any of these projects, hop to that chapter to help you through.

With one exception (that pesky but worthwhile electronic compass project, again), you can find the parts that you need to construct the projects in this chapter at any electronics store or online electronics retailer. If you don't have a well-stocked electronics outlet near you, check out both Chapter 17 and the Appendix for some mail-order electronics parts suppliers.

Unless otherwise noted, use these guidelines when selecting components:

- ✔ All resistors are rated for $1/4$ W or $1/8$ W and 5 percent or 10 percent tolerance.

- ✔ All capacitors are rated at a minimum of 25 V. We note the type of capacitor that you need (disc, electrolytic, or tantalum) in the parts list for each project.

If you want to understand the ins and outs of one or more of the electronic components you use in these projects, review the material in Chapters 3–4 and 6–8. You'll find information on resistors in Chapter 3 and a treatise on capacitors in Chapter 4. Chapter 6 explains diodes and transistors, and two of the integrated circuits (ICs) used in these projects are covered in Chapter 7. Wires, power sources, and other parts (for instance, sensors, speakers, and buzzers) are discussed in Chapter 8.

Creating Cool, Crazy, Blinky Lights

Your first mission — should you choose to accept it — is to build a circuit containing a light-emitting diode (LED) that blinks on and off. This may sound simple (and, thanks to the 555 timer IC, it is), but getting it to work means you must successfully build a complete circuit, limit the current in your circuit so it doesn't fry your LED, and set up a timer to switch the current on and off so that the light blinks. That's quite a lot for a first project!

You can see the schematic of the blinky-light project in Figure 15-1. (If you need a quick refresher course on reading schematics, flip to Chapter 10.) This figure shows you how to connect the 555 timer IC to an LED and what other parts you need to power the circuit, limit the current, and control the timing of the light. Before you actually build this circuit, you might want to do a quick analysis to understand exactly how it works.

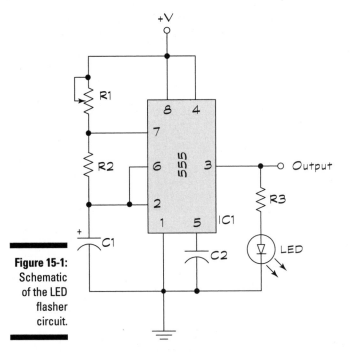

Figure 15-1:
Schematic
of the LED
flasher
circuit.

Taking a closer look at the 555 flasher

The cornerstone of this blinky-light project (as well as other projects in this chapter) is the 555 timer IC. You can use this versatile part in a variety of ways, as explained in Chapter 7. For this project, the 555 timer is configured as an *astable multivibrator,* generating an ongoing series of on/off pulses at regular intervals, sort of like an electronic metronome. The output of the 555 timer IC, at pin 3, is what you'll use to switch the LED current on and off.

Limiting current through the LED

Resistor R3 is there to keep you from frying your LED. This lowly resistor performs the important job of limiting the current passing through the LED. The output voltage at pin 3 of the 555 timer varies between 9 V (the positive power supply) when the pulse is on and 0 (zero) V when the pulse is off. Assuming the forward voltage drop across the LED is about 2.0 V (a typical value), you know that when the pulse is on, the voltage drop across resistor R3 is about 7 V. You get this result by taking the 9 V at pin 3 and subtracting the 2 V dropped across the LED. From that, you can use Ohm's Law (see Chapter 3) to calculate the current through R3, which is the same as the current through the LED, as follows:

$$\text{current} = {}^{\text{voltage}}\!/_{\text{resistance}} = {}^{7\,\text{V}}\!/_{330\,\Omega} \approx 0.021 \text{ A} = 21 \text{ mA}$$

Now, that's a current your LED can safely handle!

Controlling the timing of the pulse

As we describe in Chapter 7, both the width and the on/off timing interval of the pulse generated by the 555 timer IC are controlled by a couple of resistors (R1 and R2) and a capacitor (C1) connected to the 555 timer IC. This project uses a potentiometer to vary R1 so you can change the rate of the blinking light from slow waltz to fast samba.

If you really want to analyze and understand this circuit, review the pulse-timing equations in Chapter 7, plug in the values you use for R1 (after measuring the pot's variable resistance with your multimeter), R2, and C1; see whether the light blinks at about the rate your calculations say it should.

This circuit also makes a handy piece of test equipment. Connect the output of the 555 (pin 3 on the chip) to some other project and use this circuit as a signal source. You can see how this works in several of the other projects in this chapter that are built around the 555 chip.

Building the blinky-light circuit

It's easy to build the LED flasher circuit. Use Figure 15-2 as your guide. Note that connections to the positive side of the power supply are made along the top of the breadboard, and ground connections (the negative side of the power supply) are made along the bottom of the breadboard.

There's a bit more space between components to make it easier to see where all the parts go. You should usually build in a little bit of space, rather than squeezing things together, so you can see what you're doing.

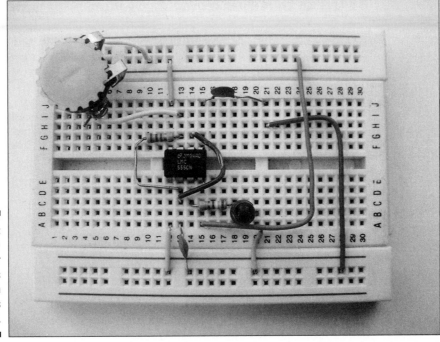

Figure 15-2:
An LED
flasher
with parts
mounted on
a solderless
board.

Running down the LED flasher parts

Here are the parts that you need to build the LED flasher circuit:

- ✔ **9 V battery** (with optional battery clip)
- ✔ **IC1:** LM555 timer IC
- ✔ **R1:** 1MΩ potentiometer
- ✔ **R2:** 47kΩ resistor
- ✔ **R3:** 330 ohm resistor
- ✔ **C1:** 1 µF tantalum (polarized) capacitor
- ✔ **C2:** 0.1 µF disc (non-polarized) capacitor
- ✔ **LED:** Light-emitting diode (any color)

Stepping through the construction of the LED flasher

Follow these steps to build the LED flasher circuit:

1. **Collect all the components you need for the project ahead of time. See the parts list below for a rundown of what you need.**

   There's nothing worse than starting a project, only to have to stop halfway through because you don't have everything at hand!

2. **Carefully insert the 555 timer chip into the middle of the board.**

It's common practice to insert an IC so that it straddles the empty middle row of the breadboard and the *clocking notch* (that little indentation or dimple on one end of the chip) faces the left of the board.

3. **Insert the two fixed resistors, R2 and R3, into the board, following the schematic and the sample breadboard in Figure 15-2.**

As noted in Chapter 7, the pins on IC chips are numbered counterclockwise, starting at the clocking notch. So, if you've placed the 555 timer IC with the clocking notch facing the left side of the board, the pin connections are as shown in Figure 15-3.

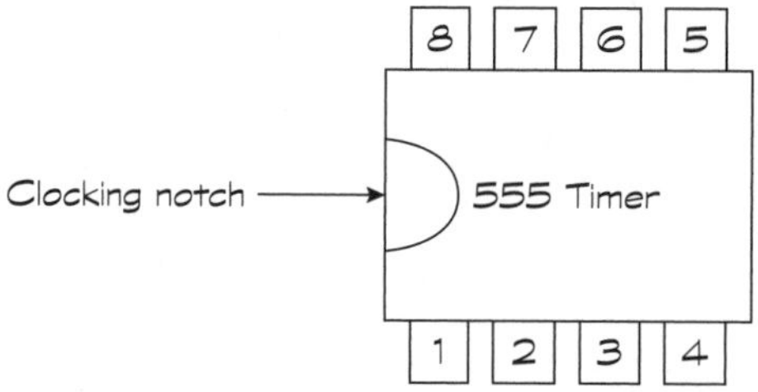

Figure 15-3: Pin connections for the 555 timer IC.

Clocking notch ——————➤ 555 Timer

8 7 6 5

1 2 3 4

4. **Insert the two capacitors, C1 and C2, into the board, following the schematic and the sample breadboard in Figure 15-2.**

Make sure you orient the polarized capacitor properly, with its negative side connected to ground.

5. **Solder wires to the potentiometer (R1) to connect it into the breadboard.**

Use 22-gauge solid strand hookup wire. The color doesn't matter. Note that the potentiometer has three connections to it. One connection (at either end) goes to pin 7 of the 555; the other two connections (at the other end and the center) are joined (or "bridged") and attached to the positive side of the power supply.

6. **Connect the LED as shown in the schematic and the photo.**

Observe proper orientation when inserting the LED: Connect the cathode (negative side, with the shorter lead) of the LED to ground. Check the packaging that came with your LED to make sure you get it right. (If you don't, and you insert the LED backwards, nothing bad will happen, but the LED won't light. Simply remove the LED, and reinsert it, the other way around.)

7. **Use 22-gauge single-strand wire, preferably already pre-cut and trimmed for use with a solderless breadboard, to finish making the connections.**

Use the sample breadboard shown in Figure 15-2 as a guide to making these jumper-wire connections.

8. *Before* **applying power, double-check your work. Verify all the proper connections by cross-checking your wiring against the schematic.**

9. **Finally, attach the 9 V battery to the positive supply and ground rails of the breadboard.**

It's easier if you use a 9 V battery clip, which contains pre-stripped leads. You may wish to solder 22-gauge solid hookup wire to the ends of the leads from the clip; this makes it easier to insert the wires into the solderless breadboard. Remember: The red lead from the battery clip is the positive terminal of the battery, and the black lead is the negative terminal, or ground.

Checking your handiwork

When you apply power to the circuit, the LED should flash. Rotate the R1 knob to change the speed of the flashing. If your circuit doesn't work, disconnect the 9 V battery and check the connections again.

Here are some common mistakes you should look for:

- ✓ **555 IC inserted backward:** This can damage the chip, so if this happens, you might want to try another 555.

- ✓ **LED inserted backward:** Pull it out and reverse the leads.

- ✓ **Connection wires and component leads not pressed firmly enough into the breadboard sockets:** Be sure that each wire fits snugly into the breadboard, so there are no loose connections.

- ✓ **Wrong component values:** Double-check, just in case!

- ✓ **Dead battery:** Try a new one.

- ✓ **Circuit wired wrong:** Have a friend take a look. Fresh eyes can catch mistakes that you might not notice.

You can use your multimeter to test voltages, currents, and resistances in your circuit. As described in Chapter 12, such tests can help you identify the cause of circuit problems. Your multimeter can tell you whether your battery has enough juice, whether your diode is still a diode, and much, much more.

If you're building a circuit that's new to you, it's good electronics practice to build it on a solderless breadboard first. That's because you often need to tweak a circuit to get it working right. When you have it working to your satisfaction, then you can make the circuit permanent if you like. Just take your time — and remember to double- (and even triple-) check your work. Don't worry — you'll be an old hand in no time.

Tapping Out a Light Tune with Piezoelectricity

Not all electronic circuits require batteries, resistors, capacitors, transistors, or any of the other components you usually find in an electronic circuit. This project consists of a neon light that glows when you tap on a piezoelectric disc, which generates its own electricity. It serves as a great demonstration of something called piezoelectricity.

Piezo — what?

The term *piezo* comes from a Greek word meaning to press or squeeze. Many years ago, some curious people with too much time on their hands found that it's possible to generate electricity by pressing certain kinds of crystals really hard. Lo and behold, these same crystals change shape — though only slightly — when you apply electricity to them. As it turns out, this was an important discovery because we use piezoelectricity in tons of everyday gadgets, such as quartz watches, alarm buzzers, guitar pickups, barbecue grill starters — and scads of other devices.

Shedding light on piezoelectricity

Figure 15-4 shows a simple circuit with a bare piezo disc and a single neon bulb. Here's a list of the (very few) parts you need to build this circuit and observe piezoelectricity in action:

- A bare piezo disc (the type that you use in a buzzer, preferably with two wires soldered on)
- Neon bulb
- Two alligator clips
- Something not-too-heavy to whack the disc with, such as a screwdriver or drumsticks (not a baseball bat)

You can find piezo discs at most electronics stores and online. Piezo discs are cheap — usually a dollar or less each — and sometimes come with two wires already soldered on. If you get one with just one wire, simply clip another wire to the edge of the disc's metal for the ground connection. Neon bulbs, sold in electronics stores, are special in that they don't light up unless you feed them at least 90 V. That's a lot of juice! But the piezo disc easily generates this much voltage.

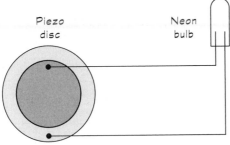

Figure 15-4:
Try this simple circuit to demonstrate the properties of piezoelectricity.

Piezo disc powering a neon bulb

To build the circuit in Figure 15-4, follow these steps:

1. **Place the disc on an insulated surface.**

   A wooden or plastic table surface works fine, but don't use a surface made of metal.

2. **Using two alligator test leads to connect the disc and the neon bulb together, as shown in Figure 15-5.**

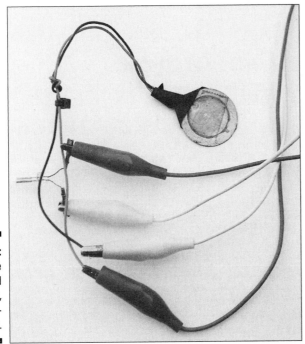

Figure 15-5:
Connect the disc and neon lamp, using alligator clips.

Place one test lead from the red wire of the disc to one connection of the neon lamp (it doesn't matter which connection). The other test lead goes from the black wire of the disc to the other connection of the neon lamp.

3. **Place the disc flat on the table.**

4. **With the plastic end of a screwdriver, rap very hard on the disc.**

   Each time you rap the disc, the neon bulb flickers.

Avoid touching the two wires that come from the disc. Although the shock you get isn't dangerous, it definitely won't feel good!

Setting up a drum line

You can build on the simple piezo-light circuit to create your very own light drum. Here's how to dazzle your audience:

1. **String up a whole slew of discs and bulbs in a row.**

2. **Tape or glue these disc-bulb combos to a plastic base.**

3. **Get a pair of drumsticks, turn down the lights, and tap on the discs in time with your favorite mood music.**

Seeing in the Dark with an Infrared Detector

Did you ever wish you could see in the dark like a cat? Now you can — by building the simple infrared detector shown in Figure 15-6. The circuit uses just three parts (plus a battery). You can make the circuit a little fancier by adding an SPST (single-pole, single-throw) switch between the positive side of the battery and the phototransistor to turn the detector on and off — or you can go the simple route and just unplug the battery when you aren't using the detector.

Detecting parts for the infrared detector

Short and sweet, here's the list of what you need to build this project:

- ✔ **9 V battery** (with optional battery clip)
- ✔ **Q1:** Infrared phototransistor (this sample circuit uses a RadioShack 276-0145, but almost any phototransistor should work fine)
- ✔ **R1:** 330 Ω resistor
- ✔ **LED:** Light-emitting diode (any color)

Be sure to use a phototransistor, and not a photodiode, in this circuit. They look the same on the outside, so check the packaging. Also, be sure to get the proper orientation for both the phototransistor and the LED. If you hook up either one backward, the circuit fails.

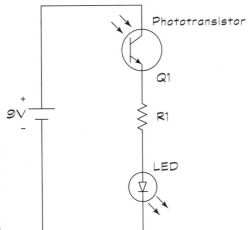

Figure 15-6:
Schematic of the infrared detector.

Sniffing out infrared light sources

Using the infrared detector, you can test for infrared light from a number of sources. Here are just two ideas to try:

- ✔ **Getting to the bottom of a remote-control dilemma:** Because remote controls use invisible infrared light, you may have a hard time figuring out what's wrong when they stop working. Does the remote have a problem, or should you blame your TV or other appliance? To test the remote control, place it up against the infrared phototransistor. Press any button on the remote; if the LED on your project flashes, you know that you have a working remote.
- ✔ **Counter-surveillance:** Check to see whether there's a hidden camera in your room. These days, covert cameras (such as the one in Figure 15-7) can "see" in the dark by using a built-in bright infrared light source. You can use your infrared detector circuit to find these infrared light sources —

even if you can't see them yourself. Turn off the lights and scan the room by holding the detector in your hand and moving it around the room. If the LED brightens, even though you don't see a light source, you may have just found the infrared light coming from a hidden camera!

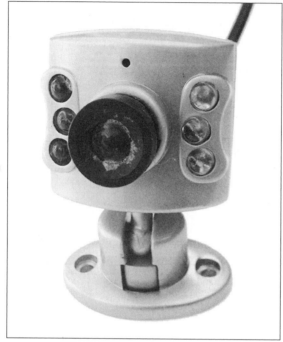

Figure 15-7:
This minia-
ture camera
can see in
the dark,
thanks to its
six infrared
light-emit-
ting diodes
(IR LEDs).

Although the infrared phototransistor is most sensitive to infrared light, it also responds to visible light. For best results, use the infrared detector in a dimly lit room. Sunlight, and direct light from desk lamps and other sources, can influence the readings.

Scaring Off the Bad Guys with a Siren

Unless you carry a badge (a real one, not the one in your toy box), you can't arrest any bad guys when you set off the warbling siren that you build in this project, shown in Figure 15-8. But it sounds cool, and you can use it as an alarm to notify you if somebody's getting at your secret stash: Baseball cards, vintage Frank Sinatra records, your signed copy of *Mister Spock's Music from Outer Space* record, or whatever.

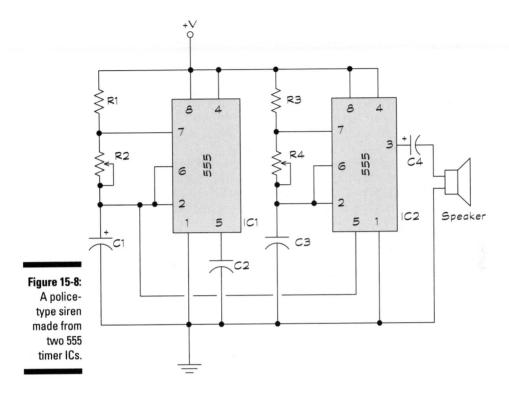

Figure 15-8:
A police-
type siren
made from
two 555
timer ICs.

Scoping out the 555 siren parts list

To start alarming your friends, gather these parts together to build the circuit:

- ✔ **9 V battery** (with optional battery clip)
- ✔ **IC1, IC2:** LM555 timer IC
- ✔ **R1, R3:** 2.2 kΩ resistors
- ✔ **R2:** 50 kΩ potentiometer
- ✔ **R4:** 100 kΩ potentiometer
- ✔ **C1:** 47 µF electrolytic (polarized) capacitor
- ✔ **C2:** 0.01 µF disc (non-polarized) capacitor
- ✔ **C3:** 0.1 µF disc (non-polarized) capacitor
- ✔ **C4:** 1 µF electrolytic or tantalum (polarized) capacitor
- ✔ **Speaker:** 8-ohm, 1 W speaker

How your warbler works

This circuit (see Figure 15-8) uses two 555 timer chips. You rig both chips to act as astable *multivibrators;* that is, they constantly change their output from low to high to low to high . . . over and over again. The two timers run at different frequencies. The timer chip on the right in the figure is configured as a *tone generator,* producing an audible frequency at its output pin (pin 3). By connecting a speaker to its output, you hear a steady, medium-pitch sound.

The timer on the left operates at a lower frequency than the timer on the right, and is used to modulate (okay, warble) the signal produced by the timer chip on the right. The signal at pin 2 of the 555 chip on the left is a slowly rising and falling ramp voltage, which you connect to pin 5 of the 555 chip on the right.

Normally, you might expect to see the signal at pin 3 of the 555 chip on the left feeding into the 555 chip on the right to trigger the second chip. As discussed in Chapter 7, pin 3 of a 555 chip is where you find the up-and-down output pulse that 555 timers are famous for. For this warbler, you get a more interesting sound by using a different signal — that at pin 2 — to trigger the second 555 chip. The signal at pin 2 of the 555 chip on the left rises and falls slowly as capacitor C1 charges and discharges. (Chapter 4 explains just how a capacitor charges and discharges; this rising and falling capacitor voltage triggers the up-and-down pulse waveform that the 555 timer outputs at pin 3, which you're not using.) By feeding this capacitor voltage (at pin 2 of the 555 chip on the left) into the control pin (pin 5) of the 555 chip on the right, you override the internal trigger circuitry of the second chip, using a varying trigger signal instead — which helps make your warbler warble.

By adjusting the two potentiometers, R2 and R4, you change the pitch and speed of the siren. You can produce all sorts of siren and other weird sound effects by adjusting these two potentiometers. You can operate this circuit at any voltage between 5 V and about 15 V. To power the gadget, use an easy-to-find 9 V battery (included in the parts list in the preceding section).

Get Lost . . . or Found, with the Electronic Compass

Discover where in the world you are with this very cool electronic compass! This magnetic compass uses the same technology that manufacturers build into many cars to show you your direction electronically. Four LEDs light up to show you the four cardinal points on the map: N (north), S (south), E (east), and W (west). The circuit illuminates adjacent LEDs to show the in-between directions, SW, SE, NW, and NE. You can see the schematic for the electronic compass in Figure 15-9.

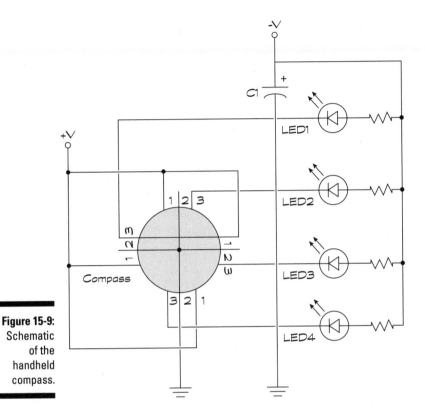

Figure 15-9:
Schematic
of the
handheld
compass.

Checking your electronic compass parts

To point you in the right direction, here are the parts that you need to gather to build your compass:

- ✔ **9 V battery**
- ✔ **COMPASS:** Dinsmore 1490 magnetic electronic compass (see the next section, "Peeking under the compass hood," for a detailed description)
- ✔ **R1-R4:** 1kΩ resistor
- ✔ **C1:** 10 µF electrolytic (polarized) capacitor
- ✔ **LED1–LED4:** Light-emitting diode (any color)
- ✔ **MISC:** Project box, switch, battery clip (all optional)

Peeking under the compass hood

The heart of this project is a special compass module, the Dinsmore 1490. This module isn't a common, everyday part. You have to special-order it, but you can have a lot of fun with the project, making it worth the $13 to $15 that you pay for the compass module. Check out the manufacturer's representative at www.robsonco.com for the compass module, and don't forget to try other possible sources by doing a Google or Yahoo! search. Try the search phrase "dinsmore compass."

The 1490 compass module is about the size of a small thimble. The bottom of the sensor has a series of 12 tiny pins, as you can see in the pinout drawing in Figure 15-10. The pins are arranged in four groups of three and consist of the following connection types:

✔ Power

✔ Ground

✔ Output (or signal)

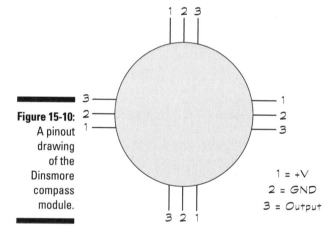

Figure 15-10: A pinout drawing of the Dinsmore compass module.

1 = +V
2 = GND
3 = Output

By doing some careful soldering, you can build a nice portable, electronic compass that you can take anywhere. Put it in a small enclosure, with the LEDs arranged in typical clockwise N, E, S, W circular orientation. You can buy enclosures at many electronics stores. Enclosures come in a variety of sizes, starting from about two inches square. Select an enclosure large enough to contain the circuit board and batteries.

You can power the compass by using a 9 V battery. Add a switch from the positive terminal of the battery to turn the unit on or off, or simply remove the battery from its clip to cut the juice and turn off your compass.

When There's Light, You Hear This Noise . . .

Figure 15-11 shows you a schematic of a light alarm. The idea of this project is simple: If a light comes on, the alarm goes off. You build the alarm around a 555 timer chip, which acts as a tone generator. When light hits the photoresistor, the change in resistance triggers transistor Q1. This response turns the 555 on, and it squeals its little heart out. You can adjust the sensitivity of the alarm by turning R1, which is a potentiometer (variable resistor).

Assembling the light alarm parts list

Here's the shopping list for the light-alarm project:

- ✔ **9 V battery** (with optional battery clip)
- ✔ **IC1:** LM555 timer IC
- ✔ **Q1:** 2N3906 PNP transistor
- ✔ **R1:** 100 kΩ potentiometer
- ✔ **R2:** 3.9 kΩ resistor
- ✔ **R3:** 10 kΩ resistor
- ✔ **R4:** 47 kΩ resistor
- ✔ **C1, C3:** 0.01 µF disc (non-polarized) capacitor
- ✔ **C2:** 1.0 µF electrolytic or tantalum (polarized) capacitor
- ✔ **Speaker:** 8-ohm, 0.5 W speaker
- ✔ **Photoresistor:** Experiment with different sizes; for example, a larger photoresistor will make the circuit a little more sensitive

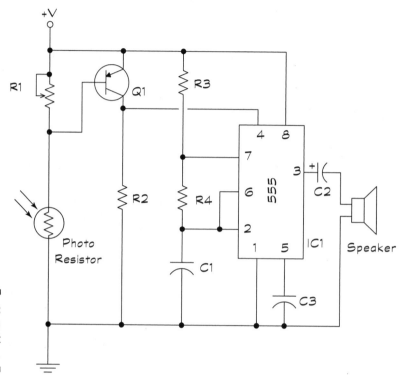

Making your alarm work for you

You can apply this handy light alarm in several practical ways. Here are just a few of 'em:

- ✔ Put the light alarm inside a pantry so it goes off whenever someone raids the chocolate chip cookies. Keep your significant other out of your stash — or keep yourself on that diet! When the pantry door opens, light comes in and the alarm goes off.

- ✔ Do you have a complex electronics project in progress in the garage that you don't want anybody to disturb? Place the alarm inside the garage, near the door. If someone opens the garage door during the day, light comes through and the alarm sounds.

- ✔ Build your own electronic rooster that wakes you up at daybreak. (Who needs an alarm clock?)

Li'l Amp, Big Sound

Give your electronics projects a big mouth with this little amplifier designed around parts that are inexpensive and easy to find at most electronics suppliers, such as the LM386 power amplifier IC. This amp boosts the volume from microphones, tone generators, and many other signal sources.

Figure 15-12 shows the schematic for this project, which consists of just six parts (including the speaker) and a battery. You can operate the amplifier at voltages between 5 V and about 15 V. A 9 V battery does the trick.

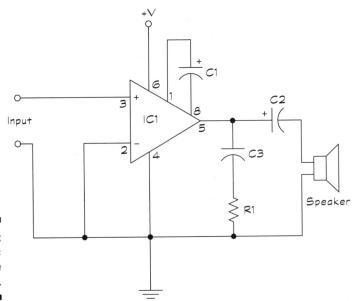

Figure 15-12:
Schematic
of the little
amplifier.

Sounding the roll call for Li'l Amp's parts

Here's a rundown of the parts that you have to gather for this project:

- ✔ **9 V battery** (with optional battery clip)
- ✔ **IC1:** LM386 power amplifier
- ✔ **R1:** 10 Ω resistor
- ✔ **C1:** 10 μF electrolytic (polarized) capacitor
- ✔ **C2:** 220 μF electrolytic (polarized) capacitor
- ✔ **C3:** 0.047 μF disc (non-polarized) capacitor
- ✔ **Speaker:** 8-ohm, 0.5 W

The better the microphone and speaker, the better the sound!

The ins and outs of Li'l Amp

To use the amplifier, connect a signal source, such as a microphone, to pin 3 of the LM386. Be sure to connect the ground of the signal source to the common ground of the amplifier circuit.

Depending on the source signal, you may find that you get better sound if you place a 0.1 μF to 10 μF capacitor between the source and pin 3 of the LM386. For smaller values (less than about 0.47 μF), use a disc capacitor; for larger values (1 μF or higher), use a tantalum capacitor. When you use a polarized capacitor, orient the positive side of the component toward the signal source.

This little amp doesn't come with a volume control, and the sound quality can take you back to your days listening to the high-school PA system. But this simple circuit puts out a whole lotta sound in a small and portable package.

Building the Handy-Dandy Water Tester

You may not be able to divine underground water with the water-tester circuit in Figure 15-13, but it can help you check for moisture in plants or find water trapped under wall-to-wall carpet.

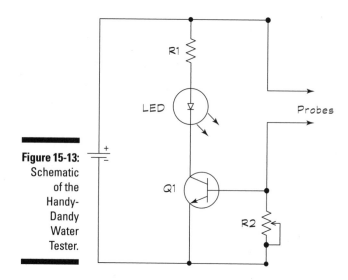

Figure 15-13: Schematic of the Handy-Dandy Water Tester.

Gathering water-tester parts

You'll need the following parts to build your water-tester project:

- ✔ **9 V battery** (with optional battery clip)
- ✔ **Q1:** 2N2222 NPN transistor
- ✔ **R1:** 470 ohm resistor
- ✔ **R2:** 50 kΩ potentiometer
- ✔ **LED:** Light-emitting diode (any color)
- ✔ **Probes:** Two small nails (4d, also called *four-penny*)

How the water tester works

The Handy-Dandy Water Tester is deceptively simple. It works by using the electrical conductivity of water (the same principle that says you don't take a bath with a plugged-in toaster in your lap). The tester contains two small metal probes. When you place the probes in water, the conductivity of the water completes a circuit. This completed circuit drives current to a transistor. When the transistor turns on, it lights an LED. When the probes aren't in contact with water (or some other conductive body), your tester has a broken (that is, open) circuit, and the LED doesn't light up.

You make the two probes with small nails, say 4d (four-penny). Place the nails about half an inch apart on a piece of plastic (but not wood or metal). The nails should be parallel to one another. File down the tips of the nails to make sharp points. These points help you drive the probes deep into the material that you're testing. For example, you can drive the probes into a carpet and pad to determine if water has seeped under the carpeting after a pipe in the next room burst.

You can adjust the sensitivity of the tester by turning potentiometer R2. Start with the potentiometer in its middle position and turn one way or another, depending on the amount of moisture or water in the object you're testing.

We suggest you use a 9 V battery, but you can power the water tester with anywhere from 5 V to 12 V.

Creating a Very Cool Lighting-Effects Generator

If you were a fan of the *Knight Rider* television series that aired back in the '80s, you remember the sequential light chaser that the Kitt Car sported in front. You can easily build your own (light-chaser setup, not car) in the garage in under an hour, using just two inexpensive ICs and a handful of other parts. The schematic for your mesmerizing lighting-effects generator is shown in Figure 15-14).

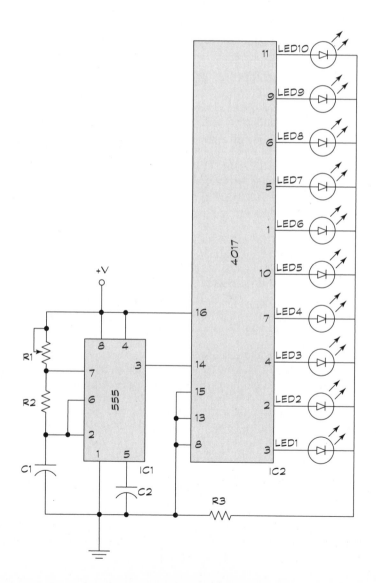

Figure 15-14: Schematic for a lighting-effects generator.

Chasing down parts for your light chaser

To start chasing lights, you need the following parts:

- **9 V battery** (with optional battery clip)
- **IC1:** LM555 timer IC
- **IC2:** 4017 CMOS decade counter IC
- **R1:** 1 MΩ potentiometer
- **R2:** 47 kΩ resistor
- **R3:** 330 ohm resistor
- **C1:** 0.47 µF disc (non-polarized) capacitor
- **C2:** 0.1 µF disc (non-polarized) capacitor
- **LED1–LED10:** Light-emitting diode (any color)

The 4017 decade counter and other CMOS chips are very sensitive to static electricity, and you can easily fry the part if you aren't careful. Make sure you take special precautions, such as wearing an anti-static wrist strap (as described in Chapter 9), before handling the 4017 CMOS IC.

Controlling the lights

The light-chaser circuit in Figure 15-14 has two sections:

- **The brains:** A 555 timer IC makes up the first section, on the left of the schematic. You wire this chip to function as an astable multivibrator. (In fact, you make the same basic circuit as the LED flasher described in the section "Creating Cool, Crazy, Blinky Lights," earlier in this chapter.) The 555 produces a series of pulses; you determine the speed of the pulses by dialing potentiometer R1.
- **The body:** The second section, on the right of the schematic, contains a 4017 CMOS decade counter chip. The 4017 chip switches each of 10 LEDs on in succession. The LEDs are switched when the 4017 receives a pulse from the 555. You wire the 4017 so it repeats the 1-to-10 sequence, over and over again, for as long as the circuit has power.

Arranging the LEDs

You can build the lighting-effects generator on a solderless breadboard just to try it out. If you plan to make it into a permanent circuit, give some thought to the arrangement of the ten LEDs. For example, to achieve different lighting effects, you can try the following:

✓ **Put all the LEDs in a row, in sequence:** The lights chase each other up (or down) over and over again.

✓ **Put all the LEDs in a row, but alternate the sequence left and right:** Wire the LEDs so the sequence starts from the outside and works its way inward.

✓ **Place the LEDs in a circle so the LEDs sequence clockwise or counter-clockwise:** This light pattern looks like a roulette wheel.

✓ **Arrange the LEDs in a heart shape:** You can use this arrangement to make a unique Valentine's Day present.

Part IV
The Part of Tens

In this part . . .

This book wouldn't be a bona fide *For Dummies* book without a handful of top 10 lists. Although this collection is a far cry from other, arguably more entertaining, top 10 lists, these contain useful tidbits that at least make for informative reading while waiting for the dentist to call you in for your root canal. (Don't you wish they were top 20 lists instead so that you would have an excuse to delay the drilling?)

In this part, we offer some top-notch tips to help you succeed as you foster your electronics habit, and we describe 10 (actually, 13) terrific sources for electronic parts, tools, and other supplies.

Chapter 16

Ten (Or So) Terrific Tips to Help You Succeed

In This Chapter

▶ Jumpstarting your electronics experience with a project kit

▶ Powering your gadgets with a variable power supply

▶ Checking the frequency of a signal

▶ Generating waveforms with a function generator

▶ Sweeping signals with a sweep generator

▶ Using a logic pulser to inject test signals into circuits

▶ Checking inputs and outputs with a logic analyzer

▶ Simulating circuits on your computer

▶ Finding great deals on testing tools

*O*kay, so you're ready to get serious about electronics, and now you're wondering what else is out there (other than basic test equipment) that can help you hone your skills and give you insight into what the heck is going on inside all those wires and components. Besides, you may be eager to impress your friends and neighbors by showing off cool gadgets you've made *right now* (not next week or next month) and to arm yourself with some impressive tools that come with lots of indicator lights, adjustable knobs, and dazzling displays. If all this is resonating with you, then you're ripe and ready to explore some of the toolkits, specialized test equipment — and even the useful (gulp) software — we describe in this chapter.

You don't absolutely, positively *need* these extras just to play around with some LEDs and resistors. A basic multimeter, and maybe a logic probe, are all you need for that. But you may want to consider the ready-made project kits described here if you'd like to jump-start your experience with circuits — and look into the additional test gear (listed a bit later) after you've gained some experience and want to graduate to bigger and better projects.

Trying Your Hand at Ready-Made Electronics Kits

If you want to make some really cool things happen with electronics, but don't want to start from scratch (and least not yet), you can purchase one of many different electronics hobby kits available from a variety of sources. These kits include everything you need to build a functional circuit: all the electronic components, wire, circuit board, and detailed instructions for putting the circuit together. Some even include an explanation of how the circuit works.

You'll find kits for light-sensitive alarms, simulated traffic signals, electronic combination locks, adjustable timers, decorative light displays, and much more. Many of the parts sources mentioned in Chapter 17 provide ready-made kits at reasonable prices. You can practice your circuit-building and analysis skills using these kits, and then move on to designing, building, and testing your own circuits from scratch.

Using a Power Supply with a Changeable Personality

You can use a *variable power supply* instead of batteries to power circuits you build and test at your workbench. A power supply produces a well-regulated (that is, very, very, *very* steady) DC voltage; most models offer voltage outputs ranging from 0 V to 20 V. The model in Figure 16-1 offers a variable output range of about 2 V to 20 V — controllable by a continuous dial — as well as preset outputs of –5 V, +5 V, and +12 V.

Figure 16-1: A variable power supply.

A power supply is characterized by its voltage range as well as its current capacity. The higher the current rating of the power supply, the heavier the load it can power. Avoid choosing a power supply with only a modest current output — say, less than 1 A. You can't adequately drive all circuits with lower currents. Instead, consider a power supply that delivers a minimum of 2 A at +5 V and at least 1 A at any other voltage.

Counting Up Those Megahertz

You can use a *frequency counter* (or frequency meter) to help you determine whether your AC circuit is operating properly. By touching the leads of a this test device to a signal point in a circuit, you can measure the frequency of that signal. For example, suppose you create an infrared transmitter and the light from that transmitter is supposed to pulse at 40,000 cycles per second (also known as 40 kHz). If you connect a frequency counter to the output of the circuit, you can verify that the circuit is indeed producing pulses at 40 kHz — not 32 kHz, 110 kHz, or some other Hz.

You can use most models, such as the one in Figure 16-2, on digital, analog, and most radio-frequency (RF) circuits (such as radio transmitters and receivers). For most hobby work, you need only a basic frequency counter; a $100 or $150 model should do just fine. In addition, some of the newer multimeters also have a rudimentary frequency-counting feature.

In digital circuits, signal voltages are limited to a range of zero up to about 12 V, but in analog circuits, voltages can vary widely. Most frequency counters are designed to work with analog voltages ranging from a few hundred millivolts to 12 V or more. Check the manual that came with your frequency counter for specifics.

Figure 16-2:
A digital frequency counter measures the frequency of a signal.

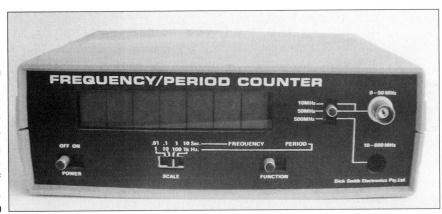

Frequency counters display the frequency signal from 0 (zero) Hz to a maximum limit based on the design of the counter. This limit usually goes well into the megahertz; it's not uncommon to find an upper limit of 25 to 50 MHz. Higher-priced frequency counters come with or offer a *prescaler* as an option — a device that extends the useful operating frequency of the frequency counter to much higher limits. Go for the prescaler feature if you're working with high-frequency radio gear or computers.

Generating All Kinds of Signals

To test a circuit's operation, it often helps to apply a known signal input to the circuit, and observe how the circuit behaves. You can use a *function generator* to create repeating-signal AC waveforms in a variety of shapes and sizes — and apply the generated waveform to the input of the circuit you're testing. Most function generators develop three kinds of waveforms: sine, triangle, and square. You can adjust the frequency of the waveforms from a low of between 0.2 Hz and 1 Hz to a high of between 2 MHz and 20 MHz. Some function generators come with a built-in frequency counter so you can accurately time the waveforms you generate. You can also use a standalone frequency counter to fine-tune the output of your function generator.

Say you're building a circuit to detect the ultrasonic pulses emitted by bats as they fly around at dusk. Most species of bats emit sound in the 20 kHz to 120 kHz frequency range (although some species emit sound at even higher frequencies) — above the range of human hearing. Bat detector circuits generally use ultrasonic transducers to convert sound within a range of frequencies into an electrical signal, and process that signal to convert it to a signal humans can detect (such as audible sound, or LED light emissions). To perform an accurate test of your bat detector circuit, you set your function generator to a frequency within the range your circuit is designed to detect and turn the amplitude way down. You then attach the two leads from your function generator to the input of your bat detector circuit, and make sure your circuit is working as designed. You can adjust the function generator frequency up and down to verify the operation of your circuit, and to make sure your circuit isn't detecting other frequencies — such as those produced by your neighbor's dog barking away.

You can find many easy-to-build bat detector electronic circuits on the Internet. Just search the Internet using the words *bat detector circuit.*

Sweeping Frequencies Up and Down

When you need to test the behavior of a circuit in response to a whole slew of input frequencies, you may need to use a sweep generator. A *sweep generator* is a type of function generator that produces a signal with a frequency that continuously varies (sweeps) within a range around a specified center frequency. Sweep generators typically vary the frequency of the output waveform within preselected limits, such as 100 Hz to 1 kHz or 1 kHz to 20 kHz, and allow you to control the sweep rate (how fast the signal frequency changes). Not only does this sweep sound like E.T. calling home (connect a speaker to the output of an audio sweep generator to hear this effect), but it also helps you identify frequency-related problems in frequency-sensitive circuits.

A *frequency-sensitive circuit* is designed to operate differently depending on the frequency of the input signal. Filter circuits, resonant circuits, and RF transmitter/receiver circuits are examples of frequency-sensitive circuits (as is the bat detector circuit described in the preceding section). If you're building a radio-receiver circuit, for example, you need to ensure it operates properly over a range of frequencies. By applying a sweep signal to the input of your circuit, you can observe (in one fell swoop) how your circuit behaves in response to a range of frequencies. You'll find a sweep generator useful in troubleshooting audio and video equipment, where altering the input frequency can reveal faulty components.

Some function generators also have a sweep feature, covering two functions with one tool.

Putting a Pulse Here, Putting a Pulse There

You can use an inexpensive handheld *logic pulser* to help you test and troubleshoot digital circuits. This pen-like device, shown in Figure 16-3, injects a high or low digital pulse into a digital circuit. (A *pulse* is simply a signal that alternates between high and low very rapidly, the way the throbbing of your arteries produces your pulse.) Many logic pulsers allow you to switch between injecting a single pulse and injecting a train of pulses at a desired frequency.

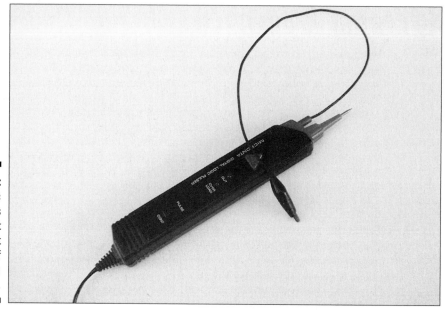

Figure 16-3:
A logic
pulser feeds
a short
signal burst
or a train of
bursts into a
circuit.

You normally use a logic pulser in conjunction with a logic probe or an oscilloscope to trace the effect of the injected pulse on your circuit. (You can read about both logic probes and oscilloscopes in Chapter 13.) For instance, you may inject a pulse into the input pin of an integrated circuit (IC) while measuring or probing the output of the IC to test whether the chip is operating properly. Logic pulsers come in handy for tracking down circuit problems because you can inject pulses into various portions of a circuit.

Most pulsers get their power from the "circuit under test" (that's a common expression referring to the circuit being tested); sometimes multiple power supplies are available. But be careful which supply you use to power the logic pulser. For instance, if you're testing a chip that's powered by 5 V, you don't want to give it a pulse powered by a 12 V supply or you'll ruin the chip. Also, some circuits work with split (+, –, and ground) power supplies, so make sure you connect the pulser's supply clips to the correct power points to avoid damage to the components.

Be sure that you don't apply a pulse to an IC pin that's designated as an output but not as an input. Some ICs are sensitive to *unloaded pulses* at their output stages, and you can destroy the chip by applying the pulse improperly. (An unloaded pulse means that the current from the pulse has no way to safely drain to another part of the circuit. If the current is applied to an output of an IC, for example, that output *could* be damaged because it's exposed to current that it's not meant to take.)

Analyzing Your Logic

To really get the lowdown on what's happening in a digital circuit, you need a logic analyzer. A *logic analyzer* is like a souped-up oscilloscope (you can read about oscilloscopes in Chapter 13): It shows you the waveforms of several inputs or outputs of a digital circuit at the same time. Many folks who are well versed in the black art of electronics find a logic analyzer much more useful than an oscilloscope for troubleshooting digital circuits.

Digital circuits often rely on the proper sequencing of signals throughout the circuit. Logic analyzers enable you to check a whole bunch of signals simultaneously. You can freeze-frame all the signals in time and observe the relationships between the signals. Then you can see if a signal is missing or doesn't time up with the others, as it should.

You can buy a stand-alone logic analyzer or one that connects to your PC. Stand-alone units cost a pretty penny, and they're fairly sophisticated. Consider getting a less expensive logic-analyzer adapter that connects to the USB, serial, or parallel ports of your computer.

You need special software that comes with the adapter. Most PC-based logic analyzers handle between 8 and 16 digital inputs at the same time.

Simulating Circuit Operation

If you've got a complicated circuit design, or just want to understand more about how a particular circuit will behave when powered up, you can use a *circuit simulator.* This software program uses computer-based models of circuit components to predict the behavior of real circuits. You tell it what components and power supplies you're using and how they should be wired up, and the software tells you whatever you want to know about the operation of the circuit: the current through any component, voltage drops across components, circuit response across various frequencies, and so forth.

Many circuit simulators are based on an industry-standard algorithm called SPICE (Simulation Program with Integrated Circuit Emphasis); you can use them to simulate and analyze various circuits — analog, digital, and *mixed-signal* (that is, incorporating both analog and digital). You can find freebie circuit simulators online, but you should know they are not guaranteed to be accurate, nor do they come with any technical support. Commercially available simulators can be expensive, but they include lots of value-added features as well as technical support. For instance, Multisym from National Instruments includes the following features (plus more):

- ✔ An extensive **model library** that contains software models of hundreds of specific manufacturer part numbers, so your simulation will show you (for example) exactly how stray capacitance from the Acme Doohickey #2 affects your sensitive circuit.

- ✔ A broad array of **waveform display tools,** including software versions of every test instrument described in this chapter (yes, your computer monitor can be made to look like an oscilloscope screen!).

- ✔ A set of in-depth **analysis tools** that help you troubleshoot your circuit and understand exactly how it will behave under various conditions, such as extreme temperature settings, or when every single component in your circuit gangs up on you by varying significantly from its nominal value (this is known as a *worst-case condition*).

- ✔ **Schematic capture tools** that allow you to control where to place the circuit symbols for your chosen circuit components on a grid displayed on your computer, how to wire-up components, and so forth, to build up a circuit diagram.

You can download a free 30-day evaluation version of Multisym and try it out yourself by visiting www.ni.com/multisim/.

Where to Get Testing-Tool Deals

We won't kid you — electronics test equipment can cost you a lot of money. Much of what you pay for is the accuracy of the device. Manufacturers strive for high accuracy so they can tout their products to businesses or meet necessary government regulations. If you're an electronics hobbyist working at home, you don't really need all that extra precision. Usually you can get by with less precise — and less expensive — models. The low-end model of a family of test products is likely to be good enough for most hobby applications, and (assuming that you take good care of it) should last many years.

You also don't need to buy everything brand new. Used and surplus items can save you a ton of cash on electronics test equipment, but there's one catch (there always is): Most of this stuff doesn't come with instruction manuals. Sometimes you can buy the manual separately or find it online. Owners of popular test gear often scan the pages of their old equipment manuals and post them online for the benefit of others.

Check out these sources for used equipment:

- ✔ **eBay and other online auction sites:** Before you bid, check the listing carefully for the condition of the equipment and the return policy if the equipment doesn't work as promised. Then check out other auctions, including those that have already ended, to see what the going price is for similar products. Set your bid accordingly, and use a proxy bidding feature so you don't have to stay glued to your computer to stay on top of the bidding. (And you'd be wise to check the seller's reviews to get an idea of the integrity of the seller you're dealing with.)

- ✔ **Electronics mail-order and local surplus outlets:** These outlets are another good source for used test equipment and are handy if you don't want to wait for an auction to finish or you prefer to know the price up front.

Whether you use an auction, mail order, or a local store, be sure that the test gear you buy actually works. Have the seller guarantee that the equipment is in working order by giving you a warranty. You may pay a little bit more for it, but if you don't make sure that it works, and you're not so good at fixing broken test gear, you may just be buying an expensive paperweight. If you're brand new to electronics, have a more experienced friend or work associate check out the gear for you.

Pass up sellers, especially on eBay or other auction sites, who aren't willing to guarantee that their products are in working order. Plenty of sellers take the time to check out their wares and guarantee that the item won't be dead on arrival.

Chapter 17

Ten (Or So) Great Electronics Parts Sources

In This Chapter

▶ Parts sources from around the globe

▶ Avoiding hazardous substances

▶ Understanding the pros and cons of surplus parts

*L*ooking for some great sources for your electronic parts? This chapter gives you some perennial favorites, both inside and outside of North America. This list is by no means exhaustive; you can find literally thousands of specialty outlets for new and used electronics. But the sources we list here are among the more established in the field, and all have Web pages for online ordering. (Some also offer print catalogs.)

North America

Check out these online resources if you're shopping within the United States or Canada. Most of these outlets ship worldwide, so if you live in a different country, you can still consider buying from these stores. Just remember that shipping costs may be higher, and you may have to pay an import duty, depending on your country's regulations.

All Electronics

www.allelectronics.com

All Electronics runs a retail store in the Los Angeles area and sends orders worldwide. Most of its stock is *new surplus,* meaning the merchandise is brand new but has been overstocked by the company. All Electronics has a printed catalog, which is also available in PDF format on its Web site. Stock changes frequently, and the latest updates are available only on the Web site. Be sure to check out the Web Only items.

Allied Electronics

www.alliedelec.com

Allied Electronics is what's known as a *stocking distributor.* It offers goods from a variety of manufacturers, and most parts are available for immediate shipping. You can also pick up your order from Allied's warehouse in Fort Worth, Texas. Allied is geared toward the electronics professional, but it welcomes hobbyists, too. The Allied catalog is *huge,* and it's available on the Web site as well as in print. You can also find parts using the search feature on Allied's Web site.

BG Micro

www.bgmicro.com

Selling primarily surplus odds-and-ends, BG Micro has great prices and terrific customer service. You can buy either from its printed catalog or online. Check the Web site for the latest deals. Stock tends to come and go quickly, so if you see something you especially like, be sure to order it now! Otherwise, some other eagle-eyed evil scientist may beat you to it.

Digi-Key

www.digikey.com

If you want it, Digi-Key probably has it. Like Allied Electronics, Digi-Key is a stocking distributor, carrying thousands upon thousands of items. Its online ordering system includes detailed product information, price, available stock levels, and even links to product datasheets. The site offers a handy search engine so you can quickly locate what you're looking for, as well as an online interactive catalog (with magnification capabilities you'll need). Digi-Key will also send a free printed catalog, but to read the tiny print, you have to get out your glasses. The text has to be teensy-weensy to fit everything in.

Electronic Goldmine

www.goldmine-elec.com

Electronic Goldmine sells new and surplus parts, from the lowly resistor to exotic lasers. Its Web site is organized by category, which makes ordering very easy. (One category, labeled "Rare and Esoteric," may be where Doc Brown got the flux capacitor that made time travel possible in *Back to the Future*). Listings for most parts include a color picture and a short description. Be sure to check out the nice selection of project kits.

Jameco Electronics

www.jameco.com

Jameco sells components, kits, tools, and more, offering both convenient online and catalog ordering. You can browse the Web site by category, or if you know the part number you're interested in — such as a 2N2222 transistor — you can find it by entering the part number into a search box. You can also use the search feature for categories of parts, such as motors, batteries, or capacitors. Just type in the category term, hit enter, and off you go.

Mouser Electronics

www.mouser.com

Similar to Allied and Digi-Key, Mouser is a stocking distributor with tens of thousands of parts on hand. If you can't find it at Mouser, it probably doesn't exist. Mouser carries more than 165,000 resistors alone, listed in the category labeled "Passive Components." You can order from its online store or its humongous print catalog. Feel free to request a printed catalog to keep on your nightstand, or let your mouse do the walking through Mouser's online catalog.

Parts Express

www.parts-express.com

Parts Express specializes in electronic parts and other equipment for audio/ visual aficionados. You'll find an ample selection of individual components, complete with user reviews, as well as project kits and other do-it-yourself resources on the Parts Express Web site. Check out the project showcase, comprehensive list of formulas (including Ohm's Law), and technical glossary, and don't forget to review the electronics safety information. You'll also find Parts Express on eBay, Facebook, and Twitter!

RadioShack

www.radioshack.com

RadioShack is perhaps the world's most recognized source for hobby electronics. You may find a RadioShack store in your favorite shopping mall, so you can shop for resistors, capacitors, underwear, and shoes all in the same trip. RadioShack also ships many of the offerings in its extensive product line by direct mail, and some of the more esoteric stuff, such as less-common integrated circuits or logic probes, is available only online.

Outside North America

Electronics is popular all over the globe! Here are some Web sites you can visit if you live in places such as Australia or the U.K. As with North American online retailers, most of these folks also ship worldwide. Check their ordering pages for details.

Dick Smith Electronics (Australia)

www.dse.com.au

Electronics from Down Under. Dick Smith Electronics offers convenient mail order (the company ships worldwide) and has more than 200 local retail stores in Australia and New Zealand.

Farnell (U.K.)

www.farnell.com

Based in the U.K., but supporting shoppers from countries worldwide, Farnell stocks some 250,000 products. You can order through Farnell's Web site.

Maplin (U.K.)

www.maplin.co.uk

Maplin provides convenient online ordering for shoppers in the U.K., Western Europe, and other international locations. The company also supports dozens of retail stores throughout the U.K.

What's RoHS Compliance?

When you're shopping around for parts, you may see the term "RoHS Compliant" next to some of the parts. The term *RoHS* (pronounced *ROW-haas*) refers to the Restriction of Hazardous Substances directive adopted in 2003 by the European Union (E.U.). The RoHS directive, which took effect in 2006, restricts the placement on the E.U. market of new electrical and electronic devices that contain more than a specified level of lead and five other hazardous substances. Companies producing consumer and industrial electronics need to worry about RoHS compliance if they want to sell products in E.U. countries (and China, which has its own RoHS specification), but if you are just tinkering around with electronics in your house, you need not worry about using lead-free solder and other RoHS-compliant parts. Just don't let your cat munch on your solder.

New or Surplus?

Surplus is a loaded word. To some, it means junk that just fills up the garage, like musty canvas tents, or funky fold-up shovels that the U.S. Army used back in the 1950s. To the true electronics buff, surplus has a totally different meaning: Affordable components that help stretch the electronics-building dollar.

Surplus just means that the original maker or buyer of the goods doesn't need it any more. It's simply excess stock for resale. In the case of electronics, surplus seldom means used, as it might for other surplus components, such as motors or mechanical devices that have been reconditioned. Except for hard-to-find components — such as older amateur radio gear — surplus electronics are typically brand new, and someone still actively manufactures much of this equipment. In this case, surplus simply means extra.

The main benefit of shopping at the surplus electronics retailer is cost: Even new components are generally lower priced than at the general electronics retailers. On the downside, you may have limited selection — whatever components the store was able to purchase. Don't expect to find every value and size of resistor or capacitor, for example.

 Remember that when you buy surplus, you don't get a manufacturer's warranty. Sometimes that's because the manufacturer is no longer in business. Although most surplus sellers accept returns if an item is defective (unless it says something different in their catalogs), you should always consider surplus stuff as-is, with no warranty implied or intended (and all that other lawyer talk).

Appendix

Internet Resources

• •

*H*ere's where you get a gaggle of interesting Internet sources for all things electronic. Businesses operate some of these sites; individuals are at the helm of the others. These lists present what we consider the most useful online resources; the idea is to save you the time and bother.

Be aware that Web sites may come and go over time. If you try to visit a site and your Web browser can't find it, the site owner probably has moved on. That's life on the Internet! Try search engines, such as Google and Yahoo!, to find additional resources.

Getting Up to Speed with Tutorials and General Information

The Web sites in this section all have worthwhile information. Browse through them to decide which sites meet your needs. The Kelsey Park School Electronics Club and the North Carolina State University Electronics Tutorial get the highest marks, but all of these sites have cool and useful information:

- ✔ **All About Circuits** (www.allaboutcircuits.com/): This site contains a series of online books on electronics. They haven't yet posted some sections, but the material they do have is well done.

- ✔ **Electronics Hobbyist** (http://amasci.com/amateur/elehob.html): Here you can find interesting articles on various basic electronics topics.

- ✔ **Graham Knott's Web Site** (http://ourworld.compuserve.com/homepages/g_knott/index1.htm): Enjoy exploring this site that Graham Knott, an electronics teacher at the University of Cambridge in England, has organized to simplify finding information on both beginning and intermediate electronics topics.

- **Kelsey Park School Electronics Club** (www.kpsec.freeuk.com): This site has a lot of good advice for newcomers to electronics projects, including a tutorial on how to read a circuit diagram, explanations of components, and a list of circuit symbols used in the U.K.

- **The North Carolina State University Electronics Tutorial** (www. courses.ncsu.edu:8020/ece480/common/htdocs): Contains good explanations of various electronics topics. Many of the illustrations are animated, which makes understanding the concepts easier.

- **Online Guide for Beginners in Electronics** (http://library.think quest.org/16497/home/index.html): Read brief introductions to several electronics topics here.

- **Williamson Labs Electronics Tutorial** (www.williamson-labs.com/home.htm): This site has some explanations of basic electronics concepts, accompanied by good illustrations that you may have fun looking through.

Figuring Things Out with Calculators

You can perform calculations on the sites in this section without having to look up equations or pick up a handheld calculator. Choose a Web site that covers the particular equation you want to use:

- **Electronics Converters and Calculators** (www.csgnetwork.com/electronicsconverters.html): This site has calculators that perform Ohm's Law calculations, parallel-resistance calculations, and resistor color-code conversions, among other handy operations.

- **The Electronics Calculator Web Site** (www.cvs1.uklinux.net/calculators/index.html): Using the tools you find on this site, you can perform calculations for Ohm's Law, RC time constants, and some other handy equations.

- **Bowen's Hobby Circuits** (www.bowdenshobbycircuits.info/): The calculators on this site include the standard calculations for Ohm's Law, RC time constants, and resistor color-code conversions. You can also find calculators for functions that you don't find on most other sites, such as a voltage-divider calculator.

Surfing for Circuits

Hungry for even more circuits to build? They're just a mouse click away! Thanks to the magic of the Internet, you can find hundreds — no, make that thousands — of electronic circuits, from basic light and sound demonstrators to advanced projects for your car or boat. Here, then, are a few of the best sites:

- **Bowden's Hobby Circuits** (www.bowdenshobbycircuits.info/): This personal site from hobbyist Bill Bowden emphasizes the why, not just the how. Here you find both circuit descriptions and alternative design suggestions.

- **Discover Circuits** (www.discovercircuits.com): This ad-supported site boasts over 28,000 schematics grouped into 500+ categories. Click on the Schematics link to view the categorized list. You can also find circuits geared for hobbyists by clicking on the Hobby Corner link.

- **Kelsey Park School Electronics Club** (www.kpsec.freeuk.com/proj.htm): You'll find some fun projects for beginners on this Web site. For each project, you'll see an explanation of how it works and a circuit layout in addition to a schematic.

Gabbing about Electronics in Discussion Forums

Use the forums on the sites in this section to get answers to your questions about projects or general electronics. Every discussion area has its own style, so spend a little time on each site to decide which forum is right for you. Post your question and others who have lived through your quandary may provide the answer you need.

We found the discussion groups on the following sites especially interesting and helpful:

- **All About Circuits Forum** (http://forum.allaboutcircuits.com): Here, you find both a general electronics discussion forum and a forum to ask for help from other forum members on any sticky projects.

- **Electronics Zone Discussion** (www.electronic-circuits-diagrams.com/forum/): This site has very active discussions on electronic circuits and projects.

- **EDAboard International Electronics Forum Center** (www.edaboard.com): Explore these active discussions about problems with projects and general electronics, along with several more specialized forums, such as one on PCB design.

- **Electronics Lab** (www.electronics-lab.com/forum/index.php): Here you can find another good site with discussions on projects, circuits, and general electronics. Check out the Project Q/A section; here, readers can post questions — and get answers — on the many projects provided in the Projects area of the site.

Be sure to take the answers you get on forums with a grain of salt. Think through the advice that you get before you build a project based solely on some well-meaning stranger's word.

Trolling for Stuff to Make Your Own Printed Circuit Boards

If you're into making your own printed circuit boards (also known as PCBs — see Chapter 11 for details), check out these Web sites for tools, chemicals, and supplies. Most of the sites in this section sell pretty much the same types of supplies, so we just list the Web pages without further description:

- **Circuit Specialists:** www.web-tronics.com
- **D&L Products:** www.dalpro.net
- **Ocean State Electronics:** www.oselectronics.com
- **Minute Man Electronics:** www.minute-man.com
- **Philmore-Datak:** www.philmore-datak.com
- **Press-n-Peel** (transfer film): www.techniks.com
- **Pulsar Professional fx** (DecalPRO transfer film): www.pulsarprofx.com

In addition to these sources, many general electronics resellers offer a limited selection of PCB-making supplies as well.

Getting Things Surplus

Looking for some good deals? Try buying surplus. Because surplus merchandise comes and goes, you have to be on your toes to catch the good stuff — but if you're lucky, you can find great bargains. Try these online surplus-electronics dealers:

- **Action Electronics** (www.action-electronics.com): This site sells both prime (brand new, direct from the manufacturer) and surplus items.
- **Alltronics** (www.alltronics.com): The inventory at this site is so huge that it may take you hours to get through it all. They have everything from used motors to teeny-tiny electronics parts.

- ✔ **American Science & Surplus** (www.sciplus.com): A trusted and reliable reseller of everything surplus. They stock some small electronics parts, but go to these guys for the motors, switches, and larger stuff.

- ✔ **Fair Radio Sales** (www.fairradio.com): This supplier has been around for about 50 years. The company specializes in ham-radio and military surplus, but it also has plenty of smaller bits and pieces to help you fill out your junk box.

- ✔ **Gateway Electronics** (www.gatewayelex.com): This site sells some kits and parts over the Internet. It also exists as a brick-and-mortar store in St. Louis, Missouri.

- ✔ **Marlin P Jones & Associates** (www.mpja.com): This site sells new and surplus electronics, test tools, and other goodies.

- ✔ **Skycraft Parts & Surplus** (www.skycraftsurplus.com): You can find a warehouse full of electronics and mechanical surplus, plus kits, test tools, and more, at this site.

In addition to these sources, be sure to also check out the top-ten list of online electronics outlets in Chapter 17.

Glossary

As with any new field of study, electronics has its own lingo. Some terms deal with electricity and units of measure, such as voltage. Other terms are labels for tools you use in projects or electronics parts, such as transistors. Here are many of the terms you'll run into throughout your electronics life. Knowing these terms will help you become electronics fluent.

alkaline battery: A type of nonrechargeable battery. See also *battery.*

alternating current (AC): Current characterized by a change in direction of the flow of electrons. See also *direct current (DC).*

ampere: The standard unit of electric current, commonly referred to as amps. One ampere is the strength of an electric current when 6.24×10^{18} electrically charged particles move past the same point within a second. See also *current, I.*

amplitude: The magnitude of an electrical signal, such as voltage or current.

anode: The terminal of a device into which conventional current (hypothetical positive charge) flows. In power-consuming devices, such as diodes, the anode is the positive terminal; in power-releasing devices, such as batteries, the anode is the negative terminal. See also *cathode.*

auto-ranging: A feature of some multimeters that automatically sets the test range.

AWG (American Wire Gauge): See *wire gauge.*

bandwidth: Relative to an oscilloscope, the highest-frequency signal that you can reliably test, measured in megahertz (MHz).

battery: A power source that uses electrochemical reaction to produce a positive voltage at one terminal and a negative voltage at the other terminal. This process involves placing two different types of metal in a certain type of chemical. See also *alkaline battery, lithium battery, nickel-cadmium battery, nickel-metal hydride battery, zinc-carbon battery.*

biasing: Applying a small amount of voltage to a diode or to the base of a transistor to establish a desired operating point.

bipolar transistor: A common type of transistor consisting of two fused pn-junctions. See also *transistor.*

breadboard: Also known as prototyping board or solderless breadboard; a rectangular plastic board (available in a variety of sizes) that contains groups of holes that little slivers of metal connect electrically. You plug in components — resistors, capacitors, diodes, transistors, and integrated circuits, for example — and then string wires to build a circuit. See also *solder breadboard.*

bus: A common connection point.

cable: A group of two or more wires protected by an outer layer of insulation, such as a common power cord.

capacitance: The capability to store energy in an electric field, measured in farads. *See also* capacitor.

capacitor: A component that provides the property of capacitance (the capability to store energy in an electric field) in a circuit. See also *capacitance.*

cathode: The terminal of a device from which conventional current (hypothetical positive charge) flows. In power-consuming devices, such as diodes, the cathode is the negative terminal; in power-releasing devices, such as batteries, the cathode is the positive terminal. See also *anode.*

circuit: A complete path that allows electric current to flow.

cladding: An extremely thin sheet of copper that you glue over a base made of plastic, epoxy, or phenol to make a printed circuit board.

closed circuit: An uninterrupted circuit through which current can flow. See also *open circuit.*

closed position: The position of a switch that allows current to flow. See also *open position.*

cold solder joint: A defective joint that occurs when solder doesn't properly flow around the metal parts.

commutator: A device used to change the direction of electric current in a motor or generator.

component: A part used in an electronics project, such as a battery or a transistor.

conductor: A substance through which electric current can move freely.

connector: A metal or plastic receptacle on a piece of equipment (a phone jack in your wall, for example) that cable ends fit into.

continuity: A type of test you perform with a multimeter to establish whether a circuit is intact between two points.

conventional current: The flow of hypothetical positive charge from positive to negative voltage; the reverse of real current. See also *real current.*

current: The flow of electrically charged particles. See also *ampere, I.*

cycle: One completion of the portion of a waveform that repeats itself. For instance, the portion of a waveform where the voltage goes from its lowest point to the highest point and back again is one cycle.

desolder pump: See *solder sucker.*

diode: A semiconductor electronic component consisting of a pn-junction that allows electric current to flow one way more easily than the other way. Diodes are commonly used to convert alternating current to direct current by limiting the flow of current to one direction.

direct current (DC): A type of current in which the electrons move in only one direction, such as the electric current generated by a battery.

double-pole, double-throw switch (DPDT): A type of switch that has two wires coming into the switch and four wires leaving the switch.

double-pole, single-throw switch (DPST): A type of switch that has two wires coming into the switch and two wires leaving the switch.

double-pole switch: A type of switch that has two input wires.

DPDT: See *double-pole, double-throw switch (DPDT).*

DPST: See *double-pole, single-throw switch (DPST).*

earth ground: A direct electrical connection to the earth. See also *ground.*

electric current: See *current.*

electricity: The movement of electrons through a conductor.

electromagnet: A temporary magnet consisting of a coiled wire around a piece of metal (typically, an iron bar). When you run current through the wire, the metal becomes magnetized. When you shut off the current, the metal loses its magnetic quality.

electromotive force: An attractive force between positive and negative charges, measured in volts.

electron: A negatively charged subatomic particle. See also *proton.*

ESD (electrostatic discharge): See *static electricity.*

fillet: A raised area formed by solder.

flathead: A term used to describe both a screw with a flat head and single slot and the screwdriver you use with it.

floating ground: A term used to describe a circuit ground that isn't connected to earth ground.

flux: A wax-like substance that helps molten solder flow around components and wire and ensures a good joint.

frequency: A measurement of how often an AC signal repeats, measured in cycles per second, or hertz (Hz). The symbol for frequency is f. See also *hertz.*

gain: The amount that a signal is amplified (the voltage of the signal coming out divided by the voltage of the signal coming in).

gauge: See *wire gauge.*

ground: A connection in a circuit used as a reference (zero volts) for a circuit. See also *earth ground.*

heat sink: A piece of metal that you attach securely to the component you want to protect. The sink draws off heat and helps prevent the heat from destroying the component.

helping hands clamp: Also sometimes called a third-hand clamp; adjustable clips that hold small parts while you're working on projects.

hertz (Hz): The measurement of the number of cycles per second in alternating current. See also *frequency.*

high signal: In digital electronics, a signal at any value higher than zero (0) volts.

I: The symbol for conventional current, measured in amperes (amps). See also *ampere, current.*

IC: See *integrated circuit (IC).*

impedance: The measure of opposition in an electrical circuit to a flow of alternating current.

inductance: The capability to store energy in a magnetic field (measured in henries). See also *inductor.*

inductor: A component that provides the property of inductance (the capability to store energy in a magnetic field) to a circuit. See also *inductance.*

infrared temperature sensor: A kind of temperature sensor that measures temperature electrically.

insulator: A substance through which electric current is unable to move freely.

integrated circuit (IC): Also known as a chip; a component that contains several miniaturized components, such as resistors, transistors, and diodes, connected in a circuit that performs a designated function.

inverter: Also known as a NOT gate; a single-input logic gate that inverts the input signal. A low input produces a high output, and a high input produces a low output. See also *logic gate.*

inverting mode: A process by which an op amp flips an input signal to produce the output signal.

jack: A type of connector. See also *connector.*

joule: A unit of energy.

lithium battery: A type of battery that generates higher voltage than other types, at about 3 volts. A lithium battery has a higher capacity than does an alkaline battery. See also *battery.*

live circuit: A circuit to which you've applied voltage.

logic gate: A digital circuit that accepts input values and determines which output value to produce based on a set of rules.

low signal: In digital electronics, a signal at or near zero (0) volts.

microcontroller: A programmable integrated circuit.

multimeter: An electronics testing device used to measure such factors as voltage, resistance, and amperage.

negative temperature coefficient (NTC) thermistor: A resistor whose resistance decreases with a rise in temperature. See also *resistor, thermistor.*

nickel-cadmium battery (NiCd): The most popular type of rechargeable battery. Some NiCad batteries exhibit the "memory effect," requiring that they be fully discharged before they can be recharged to full capacity. See also *battery.*

nickel-metal hydride battery (NiMH): A type of rechargeable battery that offers higher energy density than does a NiCad rechargeable battery. See also *battery.*

nominal value: The stated value of a resistor or other component. The real value can vary up or down from the nominal value based on the tolerance of the device. See also *tolerance.*

N-type semiconductor: A semiconductor doped with impurities so that it has more free electrons than a pure semiconductor.

ohm: A unit of resistance; its symbol is Ω. See also *R, resistance.*

Ohm's Law: An equation that defines the relationship between voltage, current, and resistance in an electrical circuit.

open circuit: A type of circuit in which a wire or component is disconnected, preventing current from flowing. See also *closed circuit.*

open position: The position of a switch that prevents current from flowing. See also *closed position.*

operational amplifier: Abbreviated as op amp; an integrated circuit containing several transistors and other components. It performs much better than an amplifier made from a single transistor. For example, an op amp can provide uniform amplification over a much wider range of frequencies than can a single-transistor amplifier.

oscillator: A circuit that generates a repeating electronic signal. See also *sine wave, square wave.*

oscilloscope: An electronic device that measures voltage, frequency, and various other parameters for varying waveforms.

pad: A contact point on a breadboard used for connecting components.

Phillips: A term used to refer to both a screw with a plus-sign-shaped (+) slot in its head and the screwdriver used with it.

piezoelectric effect: The capability of certain crystals — quartz and topaz are examples — to expand or contract when you apply voltage to them.

pn-junction: The point of contact between a P-type semiconductor, such as silicon doped with boron, and an N-type semiconductor, such as silicon doped with phosphorus. The pn-junction is the foundation for diodes and bipolar transistors. See also *bipolar transistor, diode.*

positive temperature coefficient (PTC) thermistor: A device whose resistance increases with a rise in temperature. See also *resistance, thermistor.*

potentiometer: A variable resistor that allows for the continual adjustment of resistance from virtually zero ohms to a maximum value.

power: The measure of the amount of work that electric current does while running through an electrical component, measured in watts.

precision resistor: A type of resistor with low tolerance (the allowable deviation from its stated, or nominal, value). See also *nominal value, tolerance.*

proton: A positively charged subatomic particle. See also *electron.*

prototyping board: See *breadboard.*

P-type semiconductor: A semiconductor doped with impurities so that it has fewer free electrons than a pure semiconductor.

pulse: A signal that alternates rapidly between high and low.

R: The symbol for resistance. See also *ohm, resistance.*

RC time constant: A calculation of the product of resistance and capacitance that defines the length of time it takes to charge a capacitor to two-thirds of its maximum voltage or to discharge it to one-third of its maximum voltage.

real current: The flow of electrons from a negative to a positive voltage. See also *conventional current.*

relay: A device that acts like a switch in that it closes or opens a circuit depending on the voltage supplied to it.

resistance: A measure of a component's opposition to the flow of electric current, measured in ohms. See also *ohm, R.*

resistor: A component with a fixed amount of resistance that you can add to a circuit to restrict the flow of current. See also *resistance.*

rosin flux remover: A detergent used after soldering to clean any remaining flux to prevent it from oxidizing your circuit. Available in a bottle or spray can.

schematic: A drawing showing how components in a circuit are connected.

semiconductor: A material, such as silicon, that has some of the properties of both conductors and insulators.

semiconductor temperature sensor: A kind of temperature sensor that measures temperature electrically.

sensor: An electronic component that senses a condition or an effect, such as heat or light, and converts it into an electrical signal.

series circuit: A circuit in which the current runs through each component sequentially.

short circuit: A term used to describe an accidental connection between two wires or components allowing current to pass through them rather than through the intended circuit.

sine wave: A continuous oscillatory signal defined by the mathematical sine function. See also *oscillator.*

single-pole, double-throw switch (SPDT): A type of switch that has one wire coming into the switch and two wires leaving the switch.

single-pole switch: A type of switch that has one input wire.

60/40 rosin core: Solder containing 60 percent tin and 40 percent lead (the exact ratio can vary a few percentage points) with a core of rosin flux. This type of solder is ideal for working with electronics. See also *solder, soldering.*

slide switch: A type of switch where you slide the switch forward or backward to turn something (such as a flashlight) on or off.

solar cell: A type of semiconductor that generates a current when exposed to light.

solder: A metal alloy that is heated and applied to two metal wires or leads and allowed to cool, forming a conductive joint. See also *60/40 rosin core, soldering.*

solder breadboard: A breadboard on which you solder components in place. See also *breadboard.*

soldering: The method you use in your electronics projects to assemble components on a circuit board to build a permanent electrical circuit; rather than use glue to hold things together, you use small globs of molten metal, or solder. See also *solder.*

soldering iron: A wand-like tool that consists of an insulating handle, a heating element, and a polished metal tip used to apply solder.

soldering pencil: See *soldering iron.*

solderless breadboard: See *breadboard.*

solder sucker: Also known as a desoldering pump; a tool consisting of a spring-loaded vacuum used for removing excess solder.

solder wick: Also known as solder braid; a device used to remove hard-to-reach solder. The solder wick is a flat braid of copper that works because the copper absorbs solder more easily than the tin plating of most components and printed circuit boards.

solid wire: A wire consisting of only a single strand.

SPDT: See *single-pole, double-throw switch (SPDT).*

spike: See *voltage spike.*

square wave: A repeating signal that alternates instantaneously between two different levels.

static electricity: A form of current that remains trapped in an insulating body, even after you remove the power source. Lightning is a form of static electricity.

strain relief: A device that clamps around a wire and prevents you from tugging the wire out of the enclosure.

stranded wire: Two or three small bundles of very fine wires, each wrapped in insulation.

stray capacitance: A term used to describe energy that's stored in a circuit unintentionally when electric fields occur between wires or leads that are placed too close together.

sweep generator: A type of function generator that produces a signal with a frequency that continuously varies (sweeps) within a range around a specified center frequency.

terminal: A piece of metal to which you hook up wires (as with a battery terminal).

thermistor: A resistor whose resistance value varies with changes in temperature.

thermocouple: A type of sensor that measures temperature electrically.

third-hand clamp: Also called helping hands clamp; a small, weighted clamp that holds parts while you solder.

tinning: The process of heating up a soldering tool to its full temperature and applying a small amount of solder to the tip to prevent solder from sticking to the tip.

tolerance: The allowed variation from the nominal value of a component due to the manufacturing process expressed as a percentage. See also *nominal value.*

trace: A wire on a circuit board that runs between the pads to electrically connect the components.

transistor: A semiconductor device that's commonly used to switch and amplify electrical signals.

V: The symbol for voltage, also commonly represented by E. See also *voltage.*

variable capacitor: A capacitor whose capacitance can be dynamically altered mechanically or electrically. See also *capacitance, capacitor.*

variable coil: A coil of wire surrounding a movable metal slug. By turning the slug, you change the inductance of the coil.

variable resistor: See *potentiometer.*

voltage: An attractive force between positive and negative charges.

voltage divider: A circuit that uses voltage drops to produce voltage lower than the supply voltage at specific points in the circuit.

voltage drop: The resulting lowering of voltage when voltage pulls electrons through a resistor (or other component), and the resistor absorbs some of the electrical energy.

voltage spike: A momentary rise in voltage.

watt hour: A unit of measure of energy; the ability of a device or circuit to do work.

waveform: A pattern of voltage or current fluctuations over time, which can be viewed using an oscilloscope. See also *oscilloscope.*

wire: A long strand of metal, usually made of copper, that you use in electronics projects to conduct electric current.

wire gauge: A system of measurement of the diameter of a wire.

wire wrapping: A method for connecting components on circuit boards using wire.

zinc-carbon battery: A low-quality, nonrechargeable battery. See also *battery.*

Index

• F •

● **G** ●

• N •

• U •

• V •

• *X* •

• *Z* •

Business/Accounting & Bookkeeping

Bookkeeping For Dummies
978-0-7645-9848-7

eBay Business
All-in-One For Dummies,
2nd Edition
978-0-470-38536-4

Job Interviews
For Dummies,
3rd Edition
978-0-470-17748-8

Resumes For Dummies,
5th Edition
978-0-470-08037-5

Stock Investing
For Dummies,
3rd Edition
978-0-470-40114-9

Successful Time
Management
For Dummies
978-0-470-29034-7

Computer Hardware

BlackBerry For Dummies,
3rd Edition
978-0-470-45762-7

Computers For Seniors
For Dummies
978-0-470-24055-7

iPhone For Dummies,
2nd Edition
978-0-470-42342-4

Laptops For Dummies,
3rd Edition
978-0-470-27759-1

Macs For Dummies,
10th Edition
978-0-470-27817-8

Cooking & Entertaining

Cooking Basics
For Dummies,
3rd Edition
978-0-7645-7206-7

Wine For Dummies,
4th Edition
978-0-470-04579-4

Diet & Nutrition

Dieting For Dummies,
2nd Edition
978-0-7645-4149-0

Nutrition For Dummies,
4th Edition
978-0-471-79868-2

Weight Training
For Dummies,
3rd Edition
978-0-471-76845-6

Digital Photography

Digital Photography
For Dummies,
6th Edition
978-0-470-25074-7

Photoshop Elements 7
For Dummies
978-0-470-39700-8

Gardening

Gardening Basics
For Dummies
978-0-470-03749-2

Organic Gardening
For Dummies,
2nd Edition
978-0-470-43067-5

Green/Sustainable

Green Building
& Remodeling
For Dummies
978-0-4710-17559-0

Green Cleaning
For Dummies
978-0-470-39106-8

Green IT For Dummies
978-0-470-38688-0

Health

Diabetes For Dummies,
3rd Edition
978-0-470-27086-8

Food Allergies
For Dummies
978-0-470-09584-3

Living Gluten-Free
For Dummies
978-0-471-77383-2

Hobbies/General

Chess For Dummies,
2nd Edition
978-0-7645-8404-6

Drawing For Dummies
978-0-7645-5476-6

Knitting For Dummies,
2nd Edition
978-0-470-28747-7

Organizing For Dummies
978-0-7645-5300-4

SuDoku For Dummies
978-0-470-01892-7

Home Improvement

Energy Efficient Homes
For Dummies
978-0-470-37602-7

Home Theater
For Dummies,
3rd Edition
978-0-470-41189-6

Living the Country Lifestyle
All-in-One For Dummies
978-0-470-43061-3

Solar Power Your Home
For Dummies
978-0-470-17569-9

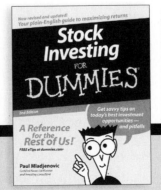

Available wherever books are sold. For more information or to order direct: U.S. customers visit www.dummies.com or call 1-877-762-2974.
U.K. customers visit www.wileyeurope.com or call (0) 1243 843291. Canadian customers visit www.wiley.ca or call 1-800-567-4797.

Internet

Blogging For Dummies,
2nd Edition
978-0-470-23017-6

eBay For Dummies,
6th Edition
978-0-470-49741-8

Facebook For Dummies
978-0-470-26273-3

Google Blogger
For Dummies
978-0-470-40742-4

Web Marketing
For Dummies,
2nd Edition
978-0-470-37181-7

WordPress For Dummies,
2nd Edition
978-0-470-40296-2

Language & Foreign Language

French For Dummies
978-0-7645-5193-2

Italian Phrases
For Dummies
978-0-7645-7203-6

Spanish For Dummies
978-0-7645-5194-9

Spanish For Dummies,
Audio Set
978-0-470-09585-0

Macintosh

Mac OS X Snow Leopard
For Dummies
978-0-470-43543-4

Math & Science

Algebra I For Dummies
978-0-7645-5325-7

Biology For Dummies
978-0-7645-5326-4

Calculus For Dummies
978-0-7645-2498-1

Chemistry For Dummies
978-0-7645-5430-8

Microsoft Office

Excel 2007 For Dummies
978-0-470-03737-9

Office 2007 All-in-One
Desk Reference
For Dummies
978-0-471-78279-7

Music

Guitar For Dummies,
2nd Edition
978-0-7645-9904-0

iPod & iTunes
For Dummies,
6th Edition
978-0-470-39062-7

Piano Exercises
For Dummies
978-0-470-38765-8

Parenting & Education

Parenting For Dummies,
2nd Edition
978-0-7645-5418-6

Type 1 Diabetes
For Dummies
978-0-470-17811-9

Pets

Cats For Dummies,
2nd Edition
978-0-7645-5275-5

Dog Training For Dummies,
2nd Edition
978-0-7645-8418-3

Puppies For Dummies,
2nd Edition
978-0-470-03717-1

Religion & Inspiration

The Bible For Dummies
978-0-7645-5296-0

Catholicism For Dummies
978-0-7645-5391-2

Women in the Bible
For Dummies
978-0-7645-8475-6

Self-Help & Relationship

Anger Management
For Dummies
978-0-470-03715-7

Overcoming Anxiety
For Dummies
978-0-7645-5447-6

Sports

Baseball For Dummies,
3rd Edition
978-0-7645-7537-2

Basketball For Dummies,
2nd Edition
978-0-7645-5248-9

Golf For Dummies,
3rd Edition
978-0-471-76871-5

Web Development

Web Design All-in-One
For Dummies
978-0-470-41796-6

Windows Vista

Windows Vista
For Dummies
978-0-471-75421-3

How-to?
How Easy.

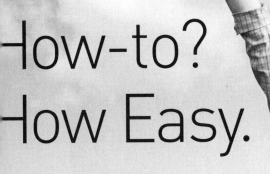

Go to www.Dummies.com

Dummies products make life easier!

DVDs • Music • Games •
DIY • Consumer Electronics •
Software • Crafts • Hobbies •
Cookware • and more!

For more information, go to
Dummies.com® and search
the store by category.

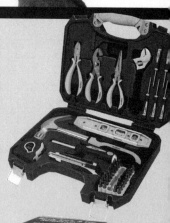